T0313748

DECENTRALIZED COVERAGE CONTROL PROBLEMS FOR MOBILE ROBOTIC SENSOR AND ACTUATOR NETWORKS

DECENTRALIZED COVERAGE CONTROL PROBLEMS FOR MOBILE ROBOTIC SENSOR AND ACTUATOR NETWORKS

ANDREY V. SAVKIN
TEDDY M. CHENG
ZHIYU XI
FAIZAN JAVED
ALEXEY S. MATVEEV
HUNG NGUYEN

IEEE Press

WILEY

Published by John Wiley & Sons, Inc., Hoboken, New Jersey. All rights reserved.
Published simultaneously in Canada.

For general information on our other products and services or for technical support, please contact our Customer Care Department within the United States at (800) 762-2974, outside the United States at (317) 572-3993 or fax (317) 572-4002.

Wiley also publishes its books in a variety of electronic formats. Some content that appears in print may not be available in electronic format. For information about Wiley products, visit our web site at www.wiley.com.

Library of Congress Cataloging-in-Publication is available.

ISBN 978-1-119-02522-1

Printed in the United States of America.

10 9 8 7 6 5 4 3

CONTENTS

PREFACE

Distributed coverage control problems for mobile robotic sensor and actuator networks have attracted considerable attention in recent years. The importance of the field of coverage control in robotic networks is quickly increasing due to the growing use of mobile robots and distributed wireless networks for sensing coverage and monitoring. The four most common types of coverage problems for mobile sensor networks are barrier coverage, sweep coverage, blanket coverage, and encircling coverage. These four types of coverage have numerous practical applications in monitoring and control of industrial and environmental processes. This book develops and studies distributed control algorithms for decentralized self-deployment of mobile robotic networks aiming to provide effective and efficient solutions of these types of coverage problems. Furthermore, for networks consisting of mobile robots that are equipped with both sensors and actuators, the problem of termination of moving environmental regions is introduced and studied.

The current book is primarily a research monograph that describes, in detail and in a unified framework, some of the latest developments on distributed coverage control problems for mobile robotic sensor and actuator networks. The intended audience of the book includes postgraduate students, researchers, and industry practitioners working in the areas of robotics, control engineering, communications, computer science, information theory, signal processing, and applied mathematics who have an interest in the emerging field of coverage control of mobile sensor/actuator networks.

This book is essentially self-contained. The reader is assumed to be competent in basic undergraduate level mathematical techniques. The theory and algorithms of distributed coverage control for mobile sensor and actuator networks are discussed in detail and illustrated by numerous examples. So a reader familiar with basic numerical methods and linear algebra will be able to understand both the algorithms presented and their mathematical analysis, as well as to implement the coverage algorithms directly from the book. The developed coverage control algorithms are fully decentralized, computationally efficient, and easily implementable in engineering practice. Furthermore, this book presents detailed mathematical studies of the proposed distributed coverage strategies.

The content of this book derives from a period of collaboration between the coauthors from 2009 to 2014. Some of the presented results have appeared in isolation in journal and conference papers. The manuscript integrates them with many new authors' original results and presents the entire material in a systematic and coherent fashion.

This book is one of the first research monographs in the diverse and challenging area of mobile robotic sensor and actuator networks. We hope that the reader finds this work both useful and interesting and is inspired to explore further the emerging field of distributed coverage control problems.

In the preparation of this work, the authors wish to acknowledge the financial support they have received from the Australian Research Council; Alexey Matveev gratefully acknowledges the support from RSF 14-21-00041. Also, the authors are grateful to the support they have received throughout the production of this book from the School of Electrical Engineering and Telecommunications at the University of New South Wales, Sydney, the Faculty of Engineering and Information Technology at the University of Technology, Sydney, and the Faculty of Mathematics and Mechanics at the Saint Petersburg State University.

Furthermore, Andrey Savkin is indebted to the endless love and support he has received from his wife Natalia, son Mikhail, and daughter Katerina. Teddy Cheng is indebted to his wife Eunice, daughter Hannah, and son Aiden for their enduring love and constant support. He also wishes to dedicate this book to his late father-in-law, Wing-Kai, in the memory of his guidance and kindness. Zhiyu Xi would like to thank her family members and colleagues for their support and tolerance throughout the years. Faizan Javed is thankful to his wife Gulrukh, their daughter Anabiya, and his parents for their continuous support and encouragement throughout this work. Alexey Matveev is grateful for the enormous support he has received from his wife Elena and daughter Julia. Hung Nguyen would like to thank his wife Lesley and his children for their wonderful support.

Andrey V. Savkin, Teddy M. Cheng, Zhiyu Xi,
Faizan Javed, Alexey S. Matveev, Hung Nguyen

CHAPTER 1

INTRODUCTION

1.1 Distributed Coverage Control of Mobile Sensor and Actuator Networks

Sensing coverage is an important issue in the field of sensor networks. For wireless sensor networks, it is considered to be a measure of their quality of service; see, e.g., [13, 32, 64, 129] and references therein. Due to the rapid development of communication and microelectronics technologies, it became possible to employ spatially distributed wireless sensor networks for performing tasks like target tracking, hazardous environment monitoring, and border surveillance in a geographically vast area. To ameliorate the performance of coverage by such a network, the sensors should be placed at proper, ideally optimal, locations. However, for a large-scale network, it would be a daunting and expensive task. To improve coverage and reduce the cost of deployment, employing movement-assisted or mobile sensors is an attractive option; see, e.g., [15, 66, 114, 122, 124]. A network of such sensors is typically implemented by deploying a group of mobile sensor equipped robots, where each robot can be viewed as a mobile sensing node. The robots can be unmanned autonomous terrain, underwater, or aerial vehicles. Their sensing capabilities are utilized to sense and monitor a spatial area of interest.

Decentralized Coverage Control Problems for Mobile Robotic Sensor and Actuator Networks.
By Andrey V. Savkin, Teddy M. Cheng, Zhiyu Xi, Faizan Javed, Alexey S. Matveev, and Hung Nguyen. Copyright © 2015 by the Institute of Electrical and Electronics Engineers, Inc.

Three types of coverage problems for mobile wireless sensor networks are defined in the seminal paper by Gage [35], namely:

1. *Barrier coverage*: to achieve a static arrangement of nodes that minimizes the probability of undetected intrusion through the barrier,

2. *Sweep coverage*: to move a number of nodes across a sensing field so that it addresses a specified balance between maximizing the detection rate of events and minimizing the number of missed detections per unit area,

3. *Blanket coverage*: to achieve a static arrangement of nodes that maximizes the detection rate of targets appearing in the sensing field.

In barrier coverage, a sensing barrier is formed by an array of sensing nodes so that any intrusion through the barrier is detected [19, 58, 59, 65, 112, 125]. Sweep coverage is achieved by moving a number of sensing nodes across a sensed field to search for and detect targets in the field [12, 26, 38, 133]. Finally, the purpose of blanket coverage is to monitor a given area so that targets appearing in this area are detected by the network of sensing nodes [29, 42, 122, 124, 137].

Practical applications of the described three types of coverage in mobile robotic sensor networks include minesweeping [12], border patrolling [59], environmental studies, detecting and localizing the sources of hazardous chemicals leakage or vapour emission, detecting sources of pollutants and plumes, environmental monitoring of disposal sites on the deep ocean floor [54], sea floor surveying for hydrocarbon exploration [9], ballistic missile tracking, bush fire monitoring, oil spill detection at high seas, environmental extremum seeking [28, 78, 137], environmental field level tracking [79], target capturing [132], and many others.

In order to achieve objectives in these coverage problems, each sensor in a network should cooperate with other sensors to fulfil a common goal. The cooperation takes place in the form of coordinated control of the sensors movement using the information from the network. However, to reduce the cost of operation, each sensor may have very limited resources, e.g., communication, sensing, or computing powers, and may suffer from severe detection and communication constraints. Therefore, the use of a centralized control algorithm is not a practical approach since not the entire global information is available to each sensor. To compensate for the lack of global information and centralized controllers, decentralized and distributed approach should be considered.

Broadly speaking, distributed control of self-deploying mobile robotic sensors falls within the general area of decentralized control, but the unique aspect of it is that mobile robotic sensors are dynamically decoupled; i.e., the motion of any sensor does not directly affect the other sensors. The study of decentralized control for groups of autonomous vehicles or robots has emerged as a challenging research area in the last decade; see, e.g., [7, 39, 45, 48, 52, 53, 69, 85, 89, 92, 96, 105, 113, 115, 127]. Distributed control laws for such groups of mobile robots are indeed motion coordination rules that rely only on a local information. In this control framework, each mobile robot is driven on the basis of information about coordinates or velocities of only a few other robots, typically its currently closest neighbors. To develop such local control rules, researchers in this new emerging area of engineering are finding much inspiration from the field of biology, where the problem of animal aggregation is central in both ecological and evolutionary theory. Animal aggregations, such as schools of fish, flocks of birds, or swarms of bees, are believed to use simple, local motion coordination rules at the individual level, which at the same time result in remarkably complex intelligent behaviors at the group level; see, e.g., [6, 34, 84, 111, 126].

To explain and simulate these behaviors, Vicsek *et al.* [120] proposed a simple discrete-time model of a system of several autonomous agents, where each agent's motion is updated using a local rule based on its own state and the states of its "neighbors". This simple but interesting model was then analytically studied by a number of researchers, e.g., [53, 55, 92, 96, 131, 135]. Moreover, modifications of the Vicsek model have also been carried out in, e.g., [55, 62, 63, 135]. For example, a Vicsek-type model with adaptive velocities is proposed in [62, 63], whereas a heterogeneous sensing Vicsek model is introduced in [130]. In addition, the converging rate of the Vicsek model is studied in [55, 135]. Also, the Vicsek model can be viewed as a special case of the model proposed in [93] for the computer animation industry to mimic animal aggregation.

To develop such local motion coordination rules, approaches like information consensus [53, 92, 96, 131] or potential field [90] are typically adopted. For low-power mobile sensor networks, consensus-based algorithms are especially attractive since they are relatively simple and require low computational cost.

Another topic of this book is study of mobile robotic sensor and actuator networks. They consist of nodes that are mobile robots (ground, underwater, or aerial unmanned vehicles). Some of these nodes are sensors and some are actuators (also called actors), whereas some nodes are endowed with both sensing and actuating capacities. Actuating capabilities are utilized to dispense control signals with the goal of achieving certain control objectives. Many modern engineering applications include the use of such networks to provide efficient and effective monitoring and control of industrial and environmental processes. These networks are able to achieve improved performance, along with reduction in power consumption and production cost. Theoretical research on mobile robotic sensors/actuator networks is at an early stage; however, some interesting theoretical results can be found in [14, 31] and references therein.

The emerging area of mobile robotic sensor and actuator networks lies at the crossroad of robotics, control engineering, computer science, and communications. The importance of this field is quickly increasing due to the growing use of wireless communications and mobile robots.

This book studies various coverage control problems for mobile sensor networks including barrier, sweep, and blanket coverage problems. Moreover, a new type of coverage referred to as encircling coverage is introduced. For mobile robotic sensor and actuator networks, the problem of termination of a moving two-dimensional region is introduced and studied. The proposed coverage control algorithms are based on the consensus approach, which was studied by many researchers in the last decade. All the robotic sensor and actuator motion algorithms developed in the book are fully decentralized and distributed, computationally efficient, easily implementable in engineering practice, and based only on information about the closest neighbors of each mobile node and local information about the environment. Moreover, the nodes have no prior information about the environment in which they operate.

It should be pointed out that this book is problem oriented, with each chapter discussing in detail distributed coverage problems and solutions that arise in the rapidly emerging area of mobile robotic sensor and actuator networks. The goal of this monograph is to present a computationally efficient, reliable, distributed, and easily implementable framework for coverage control of mobile robotic sensor and actuator networks, so that the ultimate goal of their applications (environmental or industrial monitoring, target detection and following, border protection and many others) can be fulfilled. Such a framework is very important because it is expected that future mobile wireless sensor and actuator networks will be more complex, heterogeneous, and vastly distributed. They may execute multiple tasks and consist of millions of mobile nodes.

1.2   Overview of the Book

In this section, we briefly describe the results presented in this research monograph.

Chapter 2 introduces the concept of barrier coverage and considers the problem of distributed barrier coverage between two landmarks or points. A distributed self-deployment algorithm is proposed. Moreover, we give a mathematically rigorous proof of its convergence and verify its performance by computer simulations. As always in this book, decentralized control algorithms are developed using the consensus approach, and they require only local information on the closest neighbors of each mobile robotic sensor. In this chapter, we introduce the standard Main Connectivity Assumption on the communication graph sequence of a mobile sensor network, which was first introduced in [53]. This assumption will also be used in the main results of Chapters 4, 5, 6, 7, and 8.

Chapter 3 addresses a problem of multi-level barrier coverage. The proposed distributed and decentralized control law drives a network of sensors to form K layers of parallel sensor arrays between two given points. The advantage of this law is that it is computationally efficient and easily implementable. Moreover, the sensors have no prior information about the region where the coverage is required. The main result of this chapter assumes connectivity of the communication graph of the mobile sensor network at any time, and this assumption is stronger than the Main Connectivity Assumption from Chapter 2. The main results of Chapters 9 and 10 are also based on this stronger connectivity assumption.

In Chapter 4, we theoretically develop decentralized control laws for the coordination of a mobile robotic sensor network to address the barrier and sweep coverage problems in corridor environments. The proposed control algorithms are applicable to real-time coverage operations. The control algorithms are illustrated by numerical simulations. Even though the theoretical results on the developed control algorithms are proved for straight corridors, computer simulations demonstrate that the algorithms are effective in curved corridors as well.

Chapter 5 develops a set of decentralized algorithms for the coordination of a mobile robotic sensor network to address a problem of sweep coverage along a given line. Again, the control laws are based on some consensus algorithms that are computationally efficient and easily implementable. To achieve sweep coverage, the control law drives the network of robotic mobile sensors to form a sensor barrier that moves perpendicularly to the path at a given speed. Also, the separation between each pair of the robots in the sensor barrier can be adjusted. Numerical simulations are presented for a number of scenarios to illustrate the proposed distributed algorithm. To show its practical applicability, we provide computer simulation results for an illustrative example of a sea exploration operation with multiple vessels.

Chapter 6 considers the blanket coverage problem, which is believed to be more difficult than barrier and sweep coverage problems. We propose a distributed motion coordination algorithm for a mobile sensor network to address a blanket coverage problem. Unlike the algorithms of the previous chapters, the proposed blanket coverage algorithm is randomized and has a highly probabilistic nature. According to this algorithm, the sensors are deployed on vertices of a regular triangular grid. We derive some asymptotic optimality of such grids from the classical mathematical result of Kershner [56]. We also prove convergence with probability 1 of this randomized algorithm for an arbitrary bounded region, which may be unknown to the robotic sensors a priori.

In Chapter 7, we modify the decentralized randomized control algorithm of Chapter 6 to navigate a network of autonomous mobile robotic sensors so that they collectively form a desired geometric pattern on a square grid starting from any initial deployment. In particular, we consider self-deployment with desired shapes such as interiors of a circle, an ellipse, a rectangle, a ring, and a regular hexagon. We also propose a randomized algorithm for self-deployment of a robotic sensor network in an unknown bounded region with obstacles. For the proposed distributed randomized algorithms, convergence with probability 1 is proved.

Chapter 8 introduces the problems of encircling coverage and region termination for a moving and deforming planar region and a network of mobile sensors/actuators. We propose a distributed randomized algorithm for these problems. Asymptotic optimality and convergence with probability 1 of the proposed algorithm are proved in the case of an arbitrary bounded region, which may be unknown to the mobile sensors/actuators a priori. The algorithm is decentralized, computationally efficient, and based only on information on the closest neighbors of each mobile sensor and the distance to the planar region. It should be pointed out that our algorithm for encircling and termination of a moving region is partially inspired by the famous Hannibal double-envelopment maneuver during the Battle of Cannae in which the army of Carthage under Hannibal decisively defeated a numerically superior army of the Roman Republic in 216 BC; see, e.g., [43]. Illustrative examples show that the proposed algorithm encircles and terminates moving regions in a pattern similar to encirclement and annihilation of the Roman army by the Carthaginians.

Chapter 9 addresses the blanket coverage problem that is slightly different from the blanket coverage problem studied in Chapter 6. We develop a decentralized control law that, starting from any initial deployment of the sensors, drives them to form a sensor lattice that fully covers a given two-dimensional region between two boundaries. The nodes of this lattice are in triangular pattern, which not only provides 1-coverage but also 6-connectivity. As in Chapter 6, we prove, by using the Kershner theorem, that this pattern is asymptotically optimal in terms of the minimum number of sensors required to fully cover the region. As always in this book, the proposed control law is based on the consensus algorithm approach, computationally efficient and easily implementable in practice. The control algorithm is distributed, and the control action of each sensor is based on the local information of its neighboring sensors. Moreover, the mobile sensors have no prior information about the region to be covered. It should be remarked that the blanket coverage algorithm of this chapter is totally different from the algorithm of Chapter 6. The approach of Chapter 9 is deterministic and close in spirit to the ideas of Chapters 2–5, whereas the algorithm of Chapter 6 is randomized. A number of numerical simulations is presented for different types of boundaries to illustrate the proposed algorithm.

Chapter 10 offers a decentralized formation control algorithm to coordinate a network of mobile robotic sensors so that they collectively move into a rectangular lattice pattern from any initial deployment. Numerical simulations are presented to illustrate this algorithm. Unlike blanket coverage type algorithms of Chapters 5–9 under which the sensors are eventually deployed in steady formations, the distributed algorithm of Chapter 10 ensures that the mobile robotic sensors move as a swarm with a desired geometric shape in a common direction with a common speed.

1.3   Some Other Remarks

The chapters of this book can be divided into two groups. In Chapters 2–5, 9, and 10, we propose distributed coverage algorithms that are based on building a one-dimensional structure of mobile sensors and then driving this structure to solve a barrier, sweep, or blanket coverage problem. On the other hand, Chapters 6-8 introduce a totally different class of distributed control algorithms that are based on finding consensus on a certain regular two-dimensional grid and then finding a suitable location for each mobile node on one of vertices of this grid using a randomized algorithm. Moreover, all the algorithms of Chapters 2–5, 9, and 10 are deterministic, whereas all the algorithms of Chapters 6–8 are randomized.

It should be pointed out that the mobile robotic sensor and actuator networks guided by the distributed control algorithms developed in this book can be naturally viewed as networked control systems; see, e.g., [72,74,97,98] and references therein.

Some local-rules-based intelligent behavior is expected from very large scale robotic systems. The term "very large scale robotic system" is introduced in [90] for systems consisting of autonomous robots numbering from hundreds to tens of thousands or even more. Because of decreasing costs of robots, interest in very large scale robotic systems is growing rapidly. Possible applications include underwater exploration, military surveillance, and many others. In such multi-robot systems, robots should exhibit some forms of cooperative behavior. The research presented in this monograph can be naturally viewed in the framework of very large scale robotic systems.

In this book, we use quite simple discrete-time models for motion of mobile sensors. An important direction of future research is to extend the results of the book to more realistic non-holonomic models of mobile robots. Some initial work in this direction is reported in [109], where the algorithm of Chapter 6 is applied to a problem of formation building for unicycle-like mobile robots. Such a model can describe the motion of wheeled robots, missiles, or unmanned aerial vehicles; see, e.g., [67, 75, 76, 117]. Issues such as avoidance of collisions between sensors and with obstacles, energy consumption, physical constraints of the sensors, and extensions on multidimensional problems should also be addressed in future research. A possible approach to the relevant problems of collision avoidance is to combine the algorithms from this book with known inter-robot collision avoidance and obstacle avoidance algorithms; see, e.g., [50, 51, 70, 71, 75, 99, 107, 108].

The main results of the book are originally published in [19, 21, 23–25, 100, 101, 110].

The literature in the field of distributed coverage control of mobile robotic sensor and actuator networks is vast, and we have limited ourselves to references that we found most useful or that contain material supplementing this text. The coverage of the literature in this book is by no means complete. We apologize in advance to many authors whose contributions have not been mentioned.

In conclusion, the area of mobile robotic sensor and actuator networks is a fascinating new discipline bridging robotics, control engineering, communications, information theory, computer science, and applied mathematics. The study of distributed coverage control problems represents a difficult and exciting challenge in control engineering. We hope that this research monograph will help in some small way to meet this challenge.

CHAPTER 2

BARRIER COVERAGE BETWEEN TWO LANDMARKS

2.1 Introduction

In this chapter, we address the problem of barrier coverage between two landmarks for a wireless sensor network consisting of self-deployed mobile robotic sensors. This barrier coverage problem is to deploy a group of autonomous mobile sensors to form a sensor barrier that detects any object trying to enter a protected region between two landmarks or points. A sensor barrier is defined as an array of mobile sensors such that the sensors lie on the line segment between the two landmarks in an equally spaced manner, and every point on this segment is covered by the sensing region of at least one sensor. Once a sensor barrier is formed, the sensors maintain this configuration so that any crossing paths through the line segment are covered by the sensors. The objective of this chapter is to develop a set of distributed motion coordination rules for a group of self-deployed mobile robotic sensors to achieve barrier coverage between two given landmarks. For instance, the landmarks can be given by two far-apart posts on a country's border, and the surveillance requirement is that any passing intruder has to be detected between the posts. Due to geographical vastness, it is natural to deploy a low-cost, low-power mobile wireless sensor network for such

Decentralized Coverage Control Problems for Mobile Robotic Sensor and Actuator Networks. **9**
By Andrey V. Savkin, Teddy M. Cheng, Zhiyu Xi, Faizan Javed, Alexey S. Matveev, and Hung Nguyen. Copyright © 2015 by the Institute of Electrical and Electronics Engineers, Inc.

a task. In this case, one can simply dispatch the mobile sensors near the landmarks and let them relocate themselves autonomously to meet the deployment objective.

In contrast to [59, 64, 129], we study *how to drive* a group of mobile sensors autonomously and cooperatively to address the issue of coverage in this chapter, rather than study *where* the sensors or *how many* sensors should be placed. For the sensor dispatchment problem studied in [125], a set of designated positions for the sensors is required to be precomputed and broadcasted to the sensors from a sink. Conversely, the mobile robotic sensors considered here do not have such a priori information, resulting in less computational and communication overhead. The algorithm proposed in [112] addresses a barrier coverage problem where the mobile sensors should be confined in a predefined rectangular region. In addition, a completely different approach is taken here as compared to [112, 125]: Our algorithm is developed through the use of *consensus algorithms* (see, e.g., [53, 96]) that are simple and only require local information. As a result, the algorithm is distributed and scalable.

The main results of the chapter are originally published in [19].

The remainder of the chapter is organized as follows. In Section 2.2, we formulate the problem of distributed control of mobile sensors for barrier coverage between two landmarks. Section 2.3 presents an algorithm for the barrier coverage problem and a mathematically rigorous proof of its convergence. Finally, Section 2.4 exhibits some computer simulation results to illustrate the proposed algorithm.

2.2 Problem of Barrier Coverage between Two Landmarks

In this chapter, we consider two landmarks located at L_1, $L_2 \in \mathbb{R}^2$. Let l be a unit vector such that $l = (L_2 - L_1)/\|(L_2 - L_1)\|$, where $\|\cdot\|$ denotes the Euclidean norm. The vector l characterizes the bearing of landmark L_2 relative to landmark L_1, and we write $l = [\cos(\beta) \ \sin(\beta)]^T$ for some β. Without loss of generality, we assume that $\beta \in [-\pi/2, \pi/2)$. Using l, L_1, and L_2, we define a line $\mathscr{L}_0 := \{p \in \mathbb{R}^2 : (L_2 - L_1)^T l_\perp = \mathscr{F}_0\}$, where $l_\perp$ is a normal unit vector to $\mathscr{L}_0$ such that $l^T l_\perp = 0$, and $\mathscr{F}_0$ is the associated scalar. Our problem is to develop a distributed motion coordination algorithm for a network of mobile wireless sensors to form a sensor barrier between L_1 and L_2 autonomously (see Fig. 2.1).

A mobile robotic wireless sensor network consisting of n sensors, labelled 1 through n, is considered. Given $T > 0$, the discrete-time kinematic equation of the ith sensor is

$$p_i((k+1)T) = p_i(kT) + v_i(kT)\Theta_i(kT)T, \qquad k = 0, 1, 2, \ldots. \qquad (2.1)$$

Here $p_i(\cdot) \in \mathbb{R}^2$ is the position of the sensor and $\Theta_i(\cdot) = [\cos(\theta_i(\cdot)), \sin(\theta_i(\cdot))]^T$, where $\theta_i(\cdot) \in \mathbb{R}$ is its heading measured from the x-axis in the counterclockwise direction. The velocity $v_i(\cdot)$ and the heading $\theta_i(\cdot)$ are the control inputs of sensor i, and $v_i(\cdot)$ satisfies $|v_i(t)| \leq v_{\max}$ for all $t \geq 0$ and some given $v_{\max} > 0$. At each time kT, the sensors communicate with their surrounding neighbors that are located in the disk $D_{i,r}(kT) := \{p \in \mathbb{R}^2 : \|p - p_i(kT)\| \leq r\}$, where $r > 0$ is the communication range. In addition, each mobile sensor can detect any object in the range of $s \in (0, r)$.

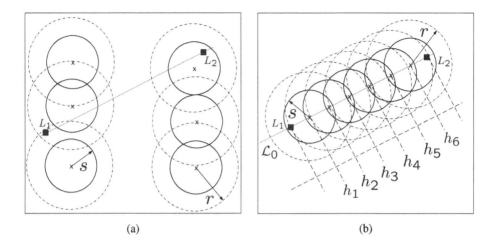

(a)   (b)

Figure 2.1   (a) Initial deployment of a group of $n = 6$ mobile sensors: There exists an undetected path between the landmarks L_1 and L_2 depicted by ■, (b) Positions of the sensors after the self-deployment: The sensors provide barrier coverage between the landmarks (r-communication range, s-sensing range).

Assumption 2.1 *The scalar $v_{\max}$ satisfies $r < v_{\max}T/\sqrt{2}$.*

Assumption 2.2 *The number of sensors, n, satisfies $(n+1)s > \|L_2 - L_1\|$.*

Assumption 2.3 *The initial headings satisfy $\theta_i(0) \in [0, \pi)$ for all $i = 1, 2, \ldots, n$.*

Remark 2.1 *For given r and $v_{\max}$, the condition $r < v_{\max}T/\sqrt{2}$ in Assumption 2.1 can be satisfied by choosing an appropriate sampling period T. Assumption 2.2 guarantees that there are sufficiently many sensors to form a sensor barrier between the landmarks. The constraint $\theta_i(0) \in [0, \pi)$ in Assumption 2.3 is introduced to avoid some "counterintuitive consequences," as is pointed out in [53].*

We introduce virtual sensors labelled 0 and $n+1$ and located at the landmarks L_1 and L_2, respectively. Let $\mathcal{N}_i(kT)$ be the set of all sensors j, $j \neq i$, and $j \in \{0, 1, 2, \ldots, n+1\}$ that at time kT belong to the disk $D_{i,r}(kT)$, and let $|\mathcal{N}_i(kT)|$ be the number of elements in $\mathcal{N}_i(kT)$. Let $\mathscr{P}$ be the collection of simple undirected graphs defined on $n+2$ vertices. For any time $kT \geq 0$, the relationship between neighbors are described by a simple undirected graph $G(kT) \in \mathscr{P}$ with vertex set $\{0, 1, 2, \ldots, n+1\}$ where i corresponds to sensor i. The vertices i and j of the graph $G(kT)$, where $i \neq j$, are connected by an edge if and only if sensors i and j are neighbors at time kT. Here, we impose the following condition on the connectivity of the graph (see [53] for more details).

Assumption 2.4 *There exists an infinite sequence of contiguous, non-empty, bounded, time-intervals $[k_j, k_{j+1})$, $j = 0, 1, 2, \ldots$, starting at $k_0 = 0$, such that across each*

$[k_j, k_{j+1})$, *the union of the collection* $\{G(kT) \in \mathscr{P} : kT \in [k_j, k_{j+1})\}$ *is a connected graph.*

This is an important assumption that will be used in several other chapters of the book. We will call Assumption 2.4 the **Main Connectivity Assumption**.

Due to the limited communication capability, the motion coordination rules for the sensors should be distributed or decentralized. For sensor i, $i = 0, 1, 2, \ldots, n, n+1$, we define $\phi_i(kT)$ as a coordination variable. For $i = 1, 2, \ldots, n$, this variable ϕ_i is initialized as $\phi_i(0) = \theta_i(0)$; whereas, for sensors 0 and $n+1$ (i.e, the landmarks), $\phi_0(kT) = \phi_{n+1}(kT) \equiv \bar{\phi}$ for all $k \geq 0$, where $\bar{\phi} = \beta + \pi/2$. The information that is available to sensor i about its neighboring sensor j is $\mathscr{I}_i^j(kT) = \{x_j(kT), y_j(kT), \phi_j(kT), j\}$.

Before stating our problem, we define n points, $h_i \in \mathbb{R}^2$ for $i = 1, 2, \ldots, n$, on the line $\mathscr{L}_0$ by $h_i := L_1 + il\|L_2 - L_1\|/(n+1)$; see Fig. 2.1(b).

Definition 2.1 (Optimal Barrier Coverage) *Given n mobile sensors and two distinct landmarks located at L_1 and L_2, a decentralized control law is said to be an optimal barrier-coverage-coordinated control for the mobile sensors between the landmarks if for almost all initial sensor positions,[1] there exists a permutation of the set $\{1, 2, \ldots, n\}$ denoted by $\{z_1, z_2, \ldots, z_n\}$ such that $\lim_{k \to \infty} \|p_{z_i}(kT) - h_i\| = 0$ for all $i = 1, 2, \ldots, n$.*

2.3   Distributed Self-Deployment Algorithm for Barrier Coverage

In this section, we develop an algorithm for the problem of barrier coverage between two landmarks. For each mobile robotic sensor i, we utilize two coordination or consensus variables, namely $\phi_i(kT)$ and $\mathscr{F}_i(kT)$. The variables characterize the line $\mathscr{L}_i(kT) := \{p \in \mathbb{R}^2 : [\cos(\phi_i(kT)) \ \sin(\phi_i(kT))]p = \mathscr{F}_i(kT)\}$ that sensor i belongs to at time kT; particularly, $\phi_i(\cdot)$ defines a vector normal to the line and $|\mathscr{F}_i(\cdot)|$ defines the minimum distance from the origin to the line. Using these variables, the aim of the algorithm is to drive the mobile sensors so that consensus is achieved with $\phi_i(kT) \to \bar{\phi}$ and $\mathscr{F}_i(kT) \to \mathscr{F}_0$; i.e., all the sensors will be on the line $\mathscr{L}_0$. At the same time, the algorithm will also distribute the sensors evenly along $\mathscr{L}_0$ between the landmarks L_1 and L_2.

To develop the algorithm for achieving consensus, at time kT and for sensor i, $i = 1, 2, \ldots, n$, we define the averages:

$$\mathscr{H}_i(kT) := \sum_{j \in \hat{\mathscr{N}}_i(kT)} \phi_j(kT)/|\hat{\mathscr{N}}_i(kT)|; \quad \mathscr{M}_i(kT) := \sum_{j \in \hat{\mathscr{N}}_i(kT)} c_{i,j}(kT)/|\hat{\mathscr{N}}_i(kT)|,$$

where $\hat{\mathscr{N}}_i(kT) := \mathscr{N}_i(kT) \cup \{i\}$ and $c_{i,j}(kT) := [\cos(\phi_i(kT)) \ \sin(\phi_i(kT))]^\top p_j(kT)$. As for the coordination variable $\mathscr{F}_i(kT)$, we define $\mathscr{F}_i(kT) := c_{i,i}(kT)$. Using $\mathscr{H}_i(kT)$

---

[1]It means for all initial conditions except for a set of zero Lebesgue measure. Lebesgue measure corresponds to the standard notion of area for subsets of $\mathbb{R}^2$; see, e.g., [94].

and $\mathcal{M}_i(kT)$, we propose the following update rules for ϕ_i and $\mathscr{F}_i$:

$$\phi_i((k+1)T) = \mathscr{H}_i(kT); \quad \mathscr{F}_i((k+1)T) = \mathcal{M}_i(kT). \tag{2.2}$$

Also, we choose $\hat{v}_i(kT) = (\mathcal{M}_i(kT) - \mathscr{F}_i(kT))/T$, i.e., the velocity component that will drive $\mathscr{F}_i(kT) \to \mathscr{F}_0$.

Next, let $q^i_j(kT)$ be the projection of the position of sensor $j \in \hat{\mathcal{N}}_i(kT)$ on the line $\mathscr{L}_i(kT)$ at time kT, i.e., $q^i_j(kT) = [\sin(\phi_i(kT)) \ -\cos(\phi_i(kT))]^\top p_j(kT)$. Define $\alpha, \beta \in \mathcal{N}_i(kT)$ such that $q^i_\alpha(kT)$ and $q^i_\beta(kT)$ are immediately next to $q^i_j(kT)$ and $q^i_\alpha(kT) < q^i_j(kT) < q^i_\beta(kT)$. Such α or β, or both, may not exist. Using $q^i_j(kT)$, $j \in \hat{\mathcal{N}}_i(kT)$, we first introduce a function $\mathcal{Q}_i(kT)$ for sensor i as follows:

$$\mathcal{Q}_i(kT) = \begin{cases} (q^i_\alpha(kT) + q^i_\beta(kT))/2 & \text{if } \alpha \text{ and } \beta \text{ exist;} \\ (q^i_i(kT) - s + q^i_\beta(kT))/2 & \text{if only } \beta \text{ exists;} \\ (q^i_\alpha(kT) + q^i_i(kT) + s)/2 & \text{if only } \alpha \text{ exists;} \\ q^i_i(kT) & \text{otherwise.} \end{cases}$$

Define $q_i(kT) := q^i_i(kT)$ and let $q_i((k+1)T)$ be the desired projected location of sensor i on the line $\mathscr{L}_i(kT)$ at time $(k+1)T$. We introduce the following update rule for $q_i((k+1)T)$:

$$q_i((k+1)T) = \begin{cases} \dfrac{q^i_i(kT) + q_0 + v_0 T}{2} & \text{if } q_0 - q^i_i(kT) \in [0, s+\varepsilon]; \\ \dfrac{q^i_i(kT) + q_{n+1} - v_0 T}{2} & \text{if } q^i_i(kT) - q_{n+1} \in [0, s+\varepsilon]; \\ \mathcal{Q}_i(kT) & \text{otherwise,} \end{cases} \tag{2.3}$$

where $q_0 := [\sin(\bar{\phi}) \ -\cos(\bar{\phi})]^\top L_1$; $q_{n+1} := [\sin(\bar{\phi}) \ -\cos(\bar{\phi})]^\top L_2$, and the scalars v_0 and ε are chosen to satisfy $0 < \varepsilon < v_0 T < (r-s)$. Using (2.3) and $q^i_i(kT)$, we define $\bar{v}_i(kT) = (q_i((k+1)T) - q^i_i(kT))/T$, i.e., the velocity component that will distribute the sensors evenly on $\mathscr{L}_0$ between L_1 and L_2.

By combining the velocity components $\bar{v}_i(kT)$ and $\hat{v}_i(kT)$, we introduce a set of decentralized control laws: For $i = 1, 2, \ldots, n$ and $k = 0, 1, 2, \ldots$, we put

$$v_i(kT) = \sqrt{\bar{v}_i(kT)^2 + \hat{v}_i(kT)^2};$$

$$\theta_i(kT) = \begin{cases} \phi_i(kT) + \xi_i(kT) - \pi/2 & \text{if } \hat{v}_i(kT) \geq 0 \\ \phi_i(kT) - \xi_i(kT) - \pi/2 & \text{if } \hat{v}_i(kT) < 0, \end{cases} \tag{2.4}$$

where $\xi_i(kT) := \cos^{-1}(\bar{v}_i(kT)/v_i(kT))$. Using (2.4), we have the following result.

Theorem 2.1 *Consider n mobile sensors described by equation (2.1). Suppose that Assumptions 2.1–2.4 are satisfied. Then the decentralized control law (2.4) is an optimal barrier-coverage coordinated control for the sensors between two given landmarks located at L_1 and L_2.*

The algorithm is summarized as follows. At time kT, each mobile sensor gathers information from its neighbors and uses it to: (1) compute $\mathcal{H}_i(kT)$, $\mathcal{F}_i(kT)$, and $\mathcal{M}_i(kT)$ in order to obtain $\hat{v}_i(kT)$; (2) determine $q_i^i(kT)$, $q_\alpha^i(kT)$, and $q_\beta^i(kT)$ (if α and/or β exist), and compute $\mathcal{Q}(kT)$ to obtain $\bar{v}_i(kT)$; and (3) generate the control input $(v_i(kT), \theta_i(kT))$ by using $\hat{v}_i(kT)$ and $\bar{v}_i(kT)$.

Proof of Theorem 2.1: Since $\phi_0(kT) = \phi_{n+1}(kT) = \bar{\phi}$ for all $k \geq 0$, Assumption 2.4 and the update law (2.2) guarantee that $\lim_{k \to \infty} \phi_i(kT) = \bar{\phi}$ and, in turn, $\lim_{k \to \infty} \mathcal{F}_i(kT) = \mathcal{F}_0$; see [53] for details. In other words, $\lim_{k \to \infty} d(p_i(kT), \mathcal{L}_0) = 0$, where $d(p_i(kT), \mathcal{L}_0)$ is the distance between sensor i and the line $\mathcal{L}_0$. Letting $d_{\max}(kT) = \max_{i=1,2,\dots,n} d(p_i(kT), \mathcal{L}_0)$, the property $\lim_{k \to \infty} d(p_i(kT), \mathcal{L}_0) = 0$ implies that for a given $\delta > 0$, there exists $\mathcal{J} \geq 0$ such that $d_{\max}(kT) < \delta$, for all $k \geq \mathcal{J}$. There also exists a permutation $\{z_1, z_2, \dots, z_n\}$ of the set $\{1, 2, \dots, n\}$ such that the projections of the positions of sensors $z_1, z_2, \dots, z_n$ on the line $\mathcal{L}_0$ satisfy

$$q_0 < q_{z_1}(kT) < q_{z_2}(kT) < \dots < q_{z_n}(kT) < q_{n+1}$$

for all $k \geq \mathcal{J}$, where $q_{z_i}(kT) := [\sin(\bar{\phi}) \ -\cos(\bar{\phi})]^T p_{z_i}(kT)$. This holds for almost all initial conditions. If $q_{z_i}(kT) = q_{z_{i+1}}(kT)$ for some i, both sensors z_i and z_{i+1} lie on the same line that is orthogonal to $\mathcal{L}_0$, and the set of initial conditions that gives rise to this situation has Lebesgue measure zero. Using (2.3), Assumptions 2.2 and 2.4 guarantee that there exists $\hat{k} \geq \mathcal{J}$ such that for all $k \geq \hat{k}$, $q_{z_1}(kT) - q_0 < s$, $q_{n+1} - q_{z_n}(kT) < s$, and $q_{z_{i+1}} - q_{z_i} < s$, $i = 1, 2, \dots, n-1$. Define the vectors $q(kT) := [q_{z_1}(kT) \ q_{z_2}(kT) \ \dots q_{z_2}(kT)]^T$ and $b := [q_0/2 \ 0 \dots 0 \ q_{n+1}/2]^T$. The update law (2.3) can be written as $q((k+1)T) = Aq(kT) + b$, where $A = A^T$. Here A is stable (see [81, p. 514]) and it can be shown that $\lim_{k \to \infty} q_{z_i}(kT) = q_0 + i \|L_2 - L_1\|/(n+1)$ for $i = 1, 2, \dots, n$. By the definition of h_i, we have $\lim_{k \to \infty} |q_{z_i}(kT) - l^T h_i| = 0$. Since $\|p_{z_i}(kT) - h_i\| \leq d(p_i(kT), \mathcal{L}_0) + |q_{z_i}(kT) - l^T h_i|$, it then gives $\lim_{k \to \infty} \|p_{z_i}(kT) - h_i\| = 0$, for $i = 1, 2, \dots, n$. This completes the proof of Theorem 2.1. ∎

2.4 Illustrative Examples

To illustrate the algorithm developed in this chapter, we deployed $n = 40$ mobile sensors for barrier coverage between two given landmarks in a simulation study, as is shown in Fig. 2.2. In the first simulation, the sensors were initially placed near the two landmarks at positions depicted by $\triangle$. By applying (2.4), they formed a sensor barrier between the landmarks after the time step $k = 2.5 \times 10^3$. In the second simulation, the sensors were initially placed near one landmark at positions depicted by $\diamond$; they formed a sensor barrier after $k = 4.5 \times 10^3$. In both cases, the sensors were evenly distributed in a sensor barrier.

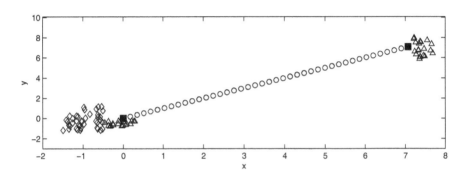

Figure 2.2 Barrier coverage between two landmarks with $n = 40$ mobile sensors: Landmarks are depicted by ■, the initial positions for case 1 by △, the initial positions for case 2 by ◇, the final positions by ○. The parameters for the simulation are chosen as $r = 0.6$, $s = 0.3$, $v_0 = 1 \times 10^{-3}$, $T = 1$, and $\varepsilon = 1 \times 10^{-5}$.

CHAPTER 3

MULTI-LEVEL BARRIER COVERAGE

3.1 Introduction

In this chapter, our focus is on multi-level barrier coverage by using a mobile robotic sensor network. Particularly, we concentrate on studying the problem of K-barrier coverage between two given points. The term K-barrier coverage was originally defined in [59]. As indicated in [35,59], barrier coverage has a wide range of applications, e.g., mine deployment, sentry duty, and border surveillance.

The K-barrier coverage problem studied in this chapter is to autonomously deploy a group of mobile robotic sensors, or simply sensors, to ultimately form K number of parallel layers of sensor barriers that detect any intruder trying to enter a protected region between two points or landmarks. These K layers form a rectangular sensor network, and any intruder's path that is orthogonal to the line segment between the points is detected by at least K distinct sensors. As shown in [65], the orthogonal crossing paths are optimal for an intruder in the sense of minimizing the probability of being detected in a two-dimensional rectangular network, provided that the intruder has no knowledge of the sensor locations.

One of the potential applications of our result is border surveillance [59]. If a country's border is a straight line, we can define the two points or landmarks as the

By Andrey V. Savkin, Teddy M. Cheng, Zhiyu Xi, Faizan Javed, Alexey S. Matveev, and Hung Nguyen. Copyright © 2015 by the Institute of Electrical and Electronics Engineers, Inc.

end points of the border, and any orthogonal path to border will be detected by K distinct sensors. If a country's border is non-straight, we can approximate the border by a number of line segments. In this case, we simply define the points as the ends of each line segment. By doing so, any path that is orthogonal to this line segment will be detected by K sensors. Whereas the previous chapter deals with only 1-barrier coverage between two points, now we are concerned with an algorithm of K-barrier coverage, where $K \geq 1$. In other words, we guarantee multiple levels of barrier coverage here as compared to only one level ensured in Chapter 2.

The main objective of this chapter is to use the information consensus approach to develop a decentralized control law for a group of autonomous mobile robotic sensors to achieve K-barrier coverage between two given points.

The main results of the chapter are originally published in [23][1].

The rest of the chapter is organized as follows. In Section 3.2, we formulate the problem of decentralized control of mobile sensors for K-barrier coverage. Section 3.3 presents a distributed algorithm for K-barrier coverage. The main result of the chapter is stated and proved in Section 3.4. Finally, Section 3.5 presents some computer simulation results to illustrate the proposed algorithm.

3.2   Problem of K-Barrier Coverage

We consider two points or landmarks in a plane and denote by L_1, $L_2 \in \mathbb{R}^2$ their locations. Our K-barrier coverage sensor deployment problem is to develop a decentralized motion coordination algorithm for a mobile robotic sensor network to cover the region between these points by K layers of sensor barriers. Initially, the mobile sensors are randomly deployed near the point L_1; see Fig. 3.1(a). To construct K layers of sensor barriers between L_1 and L_2, the mobile sensors need to move autonomously and to ultimately form a sensor lattice, as is illustrated in Fig. 3.1(b).

To state the problem, we first define a vector function $l(s)$ of $s \in \mathbb{R}$ by putting

$$l(s) = \begin{bmatrix} \cos(s) & \sin(s) \end{bmatrix}^T.$$

We also introduce the bearing of the point L_2 relative to the point L_1, i.e., the angle β such that

$$l(\beta) = (L_2 - L_1)/\|(L_2 - L_1)\| = [\cos(\beta) \quad \sin(\beta)]^T. \tag{3.1}$$

Without loss of generality, we assume that $\beta \in [-\pi/2, \pi/2)$.

Consider a mobile robotic sensor network consisting of K groups of autonomous robotic sensors, or simply sensors, where each group consists of n number of sensors. The Kn number of sensors in the network are labelled by (i, j), where $i = 1, \ldots, K$

[1]Cheng T.M. and Savkin A.V.: Self-deployment of mobile robotic sensor networks for multi-level barrier coverage. Robotica. **30(4)**, 661–669 (2012). Copyright ©2011 Cambridge University Press. Reprinted with permission.

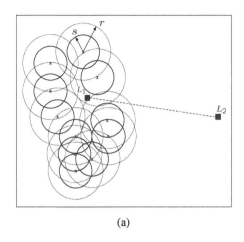
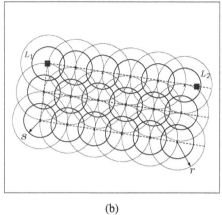

(a)   (b)

Figure 3.1   (a) Initial deployment; (b) Final deployment with $K = 3$ layers of sensor barriers.

and $j = 1,\ldots,n$. Let $v_{i,j}(\cdot)$ be the linear speed of sensor (i,j). The discrete-time kinematic equations of the sensors are given by

$$p_{(i,j)}((k+1)T) = p_{(i,j)}(kT) + Tv_{(i,j)}(kT)l(\theta_{(i,j)}(kT)). \qquad (3.2)$$

Here $p_{(i,j)}(\cdot) \in \mathbb{R}^2$ is the pair of the Cartesian coordinates of sensor (i,j), and $\theta_{(i,j)}(\cdot) \in \mathbb{R}$ is its heading measured from the x-axis in the counterclockwise direction. The speed $v_{(i,j)}$ and heading $\theta_{(i,j)}$ are the control inputs of sensor (i,j), with the speed being subjected to the constraint $|v_{(i,j)}(t)| \leq v_{\max}$.

Let $r > 0$ and $s > 0$ be the communication and sensing radii, respectively. At any discrete time instance $t = kT$, the sensors gather information about their surrounding neighbors in a range of r for coordination of their motions. In other words, sensor (i,j) communicates with its neighbors in a disk of radius r defined by

$$D_{(i,j),r}(kT) := \{p \in \mathbb{R}^2 : \|p - p_{(i,j)}(kT)\| \leq r\},$$

where $\|\cdot\|$ denotes the Euclidean norm. Also, each mobile node is equipped with a sensor that can detect objects in a range of $s \geq r/2$, i.e., objects in the following disk

$$D_{(i,j),s}(kT) := \{p \in \mathbb{R}^2 : \|p - p_{(i,j)}(kT)\| \leq s\}.$$

The communication range r of each mobile sensor satisfies $r < v_{\max}T/\sqrt{2}$, which can be met by choosing an appropriate sampling period T for given $v_{\max}$ and r.

The members of any common group composed by n sensors are called *peers*. In other words, any sensor (i,m), $m \in \{1,2,\ldots,n\} \setminus \{j\}$, is a peer of sensor (i,j). Let $\mathcal{N}_{(i,j)}(kT)$ be the set of the peers of sensor (i,j) that at time $t = kT$ belong to the disk $D_{(i,j),r}(kT)$, and let $|\mathcal{N}_{(i,j)}(kT)|$ be the number of elements in $\mathcal{N}_{(i,j)}(kT)$. These elements are called the *neighboring peers* of sensor (i,j) at time kT.

Sensor $(i,1)$ is called the *group leader* of the group $i = 1, 2, \ldots, K$. For any group leader $(i,1)$, the group leaders from $D_{(i,1),r}(kT)$ are called the *neighboring group leaders*. Let $\hat{\mathcal{N}}_{(i,1)}(kT)$ be the set of the neighboring group leaders of sensor $(i,1)$ at time $t = kT$, and let again $|\hat{\mathcal{N}}_{(i,1)}(kT)|$ be the number of elements in $\hat{\mathcal{N}}_{(i,1)}(kT)$.

To describe the relationships between neighboring peers of each group, the notion of graph is used. Let $\mathscr{P}$ be the collection of all simple undirected graphs defined on n vertices. For any time $kT \geq 0$, the relations between neighboring peers are described by a simple undirected graph $G_i(kT) \in \mathscr{P}$ with the vertex set $\{1, 2, \ldots, n\}$, where the element j corresponds to sensor (i,j). The vertices $j_1 \neq j_2$ of the graph $G_i(kT)$ are connected by an edge if and only if sensors (i, j_1) and (i, j_2) are neighboring peers at time kT.

Assumption 3.1 *The graph $G_i(kT) \in \mathscr{P}$ is connected for any $i = 1, \ldots, K$ and $k \geq 0$.*

It should be remarked that Assumption 3.1 is slightly stronger than our standard Main Connectivity Assumption (i.e., Assumption 2.4 from Chapter 2).

As for the group leaders $(1,1), (2,1), (3,1), \ldots, (K,1)$, let $\hat{\mathscr{P}}$ be the collection of all simple undirected graphs defined on K vertices. Similarly to neighboring peers, the relationships between neighboring group leaders can be described by a simple undirected graph $\hat{G}(kT) \in \hat{\mathscr{P}}$ with the vertex set $\{1, 2, \ldots, K\}$, where the element i corresponds to sensor $(i,1)$. Again, the vertices $i_1 \neq i_2$ of the graph $\hat{G}(kT)$ are connected by an edge if and only if sensors $(i_1, 1)$ and $(i_2, 1)$ are neighboring group leaders at time kT.

Assumption 3.2 *The graph $\hat{G}(kT) \in \hat{\mathscr{P}}$ is connected for any $k \geq 0$.*

Since each sensor may have a restricted communication power, the sensors only have access to local information. As a result, the control laws and coordination rules for the sensors should be distributed and decentralized in the sense that the movement of each sensor relies on only local data. These data may be about some coordination variables available to the neighboring peers and group leaders.

For sensor (i,j), we introduce the coordination variable $\phi_{(i,j)}(\cdot)$. Its current value $\phi_{(i,j)}(kT)$ is communicated to any neighboring sensor of sensor (i,j) that is in the disk $D_{(i,j),r}(kT)$. For sensors $(i,j) \neq (1,1)$, this variable is initialized as follows

$$\phi_{(i,j)}(0) = \theta_{(i,j)}(0). \tag{3.3}$$

For sensor $(1,1)$, its coordination variable $\phi_{(1,1)}(\cdot)$ is constant and given by

$$\phi_{(1,1)}(kT) \equiv \bar{\phi}, \quad k = 0, 1, 2, \ldots, \tag{3.4}$$

where $\bar{\phi} := \beta + \pi/2$. Also, sensor $(1,1)$ is placed at L_1 to indicate where the sensor barriers should be formed. We assume that $\theta_{(i,j)}(0) \in [0, \pi)$.

As was discussed, each sensor relies on only local information. Denoting by $\mathscr{I}_{(i,j)}^{(i_1,j_1)}(kT)$ the information that is available to sensor (i,j) about sensor (i_1, j_1) at time kT, we assume that it is void if sensor (i_1, j_1) is not in the disk $D_{(i,j),r}(kT)$ at this time; otherwise,

$$\mathscr{I}_{(i,j)}^{(i_1,j_1)}(kT) = \{p_{(i_1,j_1)}(kT), \phi_{i_1,j_1}(kT), (i_1,j_1)\}.$$

Let $d_1 < r$ be a desired distance between neighboring peers. We assume that there are sufficiently many sensors to form sensor barriers between the points L_1 and L_2: the number n of sensors in each group satisfies

$$(n+1)d_1 > \|L_2 - L_1\|. \tag{3.5}$$

Let $d_2 < r$ be a desired distance between neighboring group leaders.

To complete the problem setup, we introduce K lines that are parallel to the unit vector $l(\beta)$ defined in (3.1). Specifically, for $i = 1, 2, \ldots, K$, we put

$$\mathscr{L}_i := \{p \in \mathbb{R}^2 : l(\bar{\phi})^T p = \mathscr{G}_i\}, \tag{3.6}$$

where

$$l(\bar{\phi}) := [\cos(\bar{\phi}) \ \sin(\bar{\phi})]^T, \quad \mathscr{G}_1 := l(\bar{\phi})^T L_1, \quad \mathscr{G}_i := \mathscr{G}_1 - (i-1)d_2,$$

and $l(\bar{\phi})^T l(\beta) = 0$ by the definition of $\bar{\phi}$. Next, we define a line $\mathscr{W}_1$ that is perpendicular to $\mathscr{L}_i$, $i = 1, 2, \ldots, K$ by putting

$$\mathscr{W}_1 := \{p \in \mathbb{R}^2 : l(\beta)^T (p - L_1) = 0\}. \tag{3.7}$$

Using the lines $\mathscr{L}_1, \ldots, \mathscr{L}_K$, we define Kn number of locations at which the sensors should be placed. Note that these locations are *unknown* to the sensors. To define these locations, we start with introducing K number of points $a_i \in \mathbb{R}^2$ such that

$$a_1 := L_1, \quad a_i := \mathscr{L}_i \cap \mathscr{W}_1 \tag{3.8}$$

for $i = 2, 3, \ldots, K$. The above desired locations are given by

$$h_{i,j} := a_i + d_1(j-1)l(\beta) \tag{3.9}$$

for $i = 1, 2, \ldots, K$ and $j = 1, 2, \ldots, n$. The locations of $h_{i,j}$ are illustrated in Fig. 3.2, where $K = 3$ and $n = 5$.

The objective of this chapter is to design a family of distributed control laws that steers the sensors so that they converge to the points $h_{i,j}$.

Definition 3.1 (K-Barrier Coverage) *Given Kn mobile robotic sensors and two distinct points L_1 and L_2, a set of distributed motion control laws is said to be a K-barrier coverage coordinated control for the sensors between the points L_1 and L_2 if for almost all initial sensor positions, there exists a permutation of the set $\{1, 2, \ldots, K\}$ denoted by $\{x_1, x_2, \ldots, x_K\}$ such that*

$$\lim_{k \to \infty} \|p_{(x_i,1)}(kT) - h_{i,1}\| = 0, \quad i = 1, 2, \ldots, K; \tag{3.10}$$

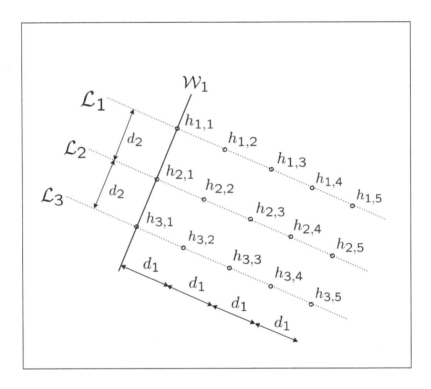

Figure 3.2   Desired locations of mobile sensors for K-barrier coverage ($K = 3$, $n = 5$).

and for each group x_i, there exists a permutation $\{y_2^{(x_i)}, y_3^{(x_i)}, \ldots, y_n^{(x_i)}\}$ of the set $\{2, 3, \ldots, n\}$ such that

$$\lim_{k \to \infty} \|P_{(x_i, y_j^{(x_i)})}(kT) - h_{i,j}\| = 0, \quad j = 2, 3, \ldots, n. \tag{3.11}$$

We recall that "for almost all" means "for all except for a set of zero Lebesgue measure".

3.3 Distributed Algorithm for K-Barrier Coverage

In this section, a set of decentralized control laws is developed for the coordination of the sensors to achieve K-barrier coverage.

At time $t = kT$ sensor $(i, 1)$ with $i = 2, 3 \ldots, K$ computes the average of the coordination variables $\phi_{(i,j)}$ from the set $\hat{\mathcal{N}}_{(i,1)}(kT)$ as follows:

$$\mathcal{A}_{(i,1)}(kT) := \frac{1}{1 + |\hat{\mathcal{N}}_{(i,1)}(kT)|} \left(\phi_{(i,1)}(kT) + \sum_{\bar{i} \in \hat{\mathcal{N}}_{(i,1)}(kT)} \phi_{(\bar{i},1)}(kT) \right). \tag{3.12}$$

Somewhat similar averages are computed by the other sensors (i,j) $(i=1,2,\ldots,K,$ $j=2,3,\ldots,n)$:

$$\mathscr{A}_{(i,j)}(kT) := \frac{1}{1+|\mathscr{N}_{(i,j)}(kT)|}\left(\phi_{(i,j)}(kT) + \sum_{\bar{j}\in\mathscr{N}_{(i,j)}(kT)}\phi_{(i,\bar{j})}(kT)\right). \qquad (3.13)$$

For each neighboring peer of such sensor (i,j), we define the following scalars:

$$c_{(\bar{i},\bar{j})}(kT) := l(\phi_{(i,j)}(kT))^T p_{(\bar{i},\bar{j})}(kT), \qquad (3.14)$$

where $(\bar{i},\bar{j})\in\mathscr{N}_{(i,j)}(kT)$. On the other hand, for each neighboring group leader $(\bar{i},1)\in\hat{\mathscr{N}}_{(i,1)}(kT)$ of sensor $(i,1)$ with $i=2,3,\ldots,K$, we define the following scalar:

$$c_{(\bar{i},1)}(kT) = l(\psi_{(i,1)}(kT))^T p_{(\bar{i},1)}(kT), \qquad (3.15)$$

where $\psi_{(i,1)}(kT) := \phi_{(i,1)}(kT) - \pi/2$.

For each sensor (i,j), we introduce a scalar $\mathscr{F}_{(i,j)}(kT)$ by putting

$$\mathscr{F}_{(i,j)}(kT) = \begin{cases} l(\phi_{(i,j)}(kT))^T p_{(i,j)}(kT) & \text{for } j=2,3,\ldots,n, \\ l(\psi_{(i,j)}(kT))^T p_{(i,j)}(kT) & \text{for } j=1, \end{cases} \qquad (3.16)$$

where $i=1,2,\ldots,K$ and $\psi_{(i,j)}(kT) = \phi_{(i,j)}(kT) - \pi/2$.

For each sensor (i,j), we also define a scalar $\mathscr{M}_{i,j}$ similarly to (3.12) and (3.13): For sensor $(i,1)$, $i=1,2,\ldots,K$, we put

$$\mathscr{M}_{(i,1)}(kT) := \frac{1}{1+|\hat{\mathscr{N}}_{(i,1)}(kT)|}\left(\mathscr{F}_{(i,1)}(kT) + \sum_{\bar{i}\in\hat{\mathscr{N}}_{(i,1)}(kT)}c_{(\bar{i},1)}(kT)\right) \qquad (3.17)$$

and for sensor (i,j), $i=1,2,\ldots,K$, $j=2,3,\ldots,n$, we put

$$\mathscr{M}_{(i,j)}(kT) := \frac{1}{1+|\mathscr{N}_{(i,j)}(kT)|}\left(\mathscr{F}_{(i,j)}(kT) + \sum_{\bar{j}\in\mathscr{N}_{(i,j)}(kT)}c_{(i,\bar{j})}(kT)\right). \qquad (3.18)$$

Using $\mathscr{A}_{(i,j)}(kT)$ and $\mathscr{M}_{(i,j)}(kT)$, we introduce $\hat{\mathscr{A}}_{i,j}(kT)$ and $\hat{\mathscr{U}}_{i,j}(kT)$ as follows:

$$\hat{\mathscr{A}}_{(i,j)}(kT) = \begin{cases} \bar{\phi} & \text{if } i=j=1, \\ \mathscr{A}_{(i,j)}(kT) & \text{otherwise}; \end{cases}$$

$$\hat{\mathscr{U}}_{(i,j)}(kT) = \begin{cases} \mathscr{F}_{(i,j)}(0) & \text{if } i=j=1, \\ \mathscr{M}_{(i,j)}(kT) & \text{otherwise}. \end{cases} \qquad (3.19)$$

Next, we define a line $\mathscr{L}_{(i,j)}(kT)$ for sensor (i,j) as follows:

if $j=1$,

$$\mathscr{L}_{(i,1)}(kT) = \{p\in\mathbb{R}^2 : l(\psi_{(i,1)}(kT))^T p = \mathscr{M}_{(i,1)}(kT)\};$$

if $j=2,3,\ldots,n$,

$$\mathscr{L}_{(i,j)}(kT) = \{p\in\mathbb{R}^2 : l(\phi_{(i,j)}(kT))^T p = \mathscr{M}_{(i,j)}(kT)\}. \qquad (3.20)$$

Sensor (i,j) belongs to the line $\mathcal{L}_{(i,j)}(kT)$, and this line thus features sensor (i,j).

Let $q^{(i,j)}_{(\bar{i},\bar{j})}(kT)$ be the projection of the position of sensor $(\bar{i},\bar{j})$ on the line $\mathcal{L}_{(i,j)}(kT)$ at time kT:

$$\text{if } j=1 \text{ and } (\bar{i},1) \in \hat{\mathcal{N}}_{(i,1)}(kT),$$
$$q^{(i,1)}_{(\bar{i},1)}(kT) = l(\psi_{(\bar{i},1)}(kT) - \pi/2)^T p_{(\bar{i},1)}(kT);$$
$$\text{if } j=2,3,\ldots,n \text{ and } (\bar{i},\bar{j}) \in \mathcal{N}_{(i,j)}(kT),$$
$$q^{(i,j)}_{(\bar{i},\bar{j})}(kT) = l(\phi_{(\bar{i},\bar{j})}(kT) - \pi/2)^T p_{(\bar{i},\bar{j})}(kT). \tag{3.21}$$

For sensor (i,j), we define the indices $\alpha, \beta \in \mathcal{N}_i(kT)$ (for $j \neq 1$) or $\alpha, \beta \in \hat{\mathcal{N}}_i(kT)$ (for $j=1$), if they exist, such that

$$q^{(i,j)}_\alpha(kT) < q^{(i,j)}_{(i,j)}(kT) < q^{(i,j)}_\beta(kT).$$

Using $q^{(i,j)}_{(i,j)}(kT)$, $q^{(i,j)}_\alpha(kT)$, and $q^{(i,j)}_\beta(kT)$, we introduce the function

$$\mathcal{D}_{(i,j)}(kT) = \begin{cases} \dfrac{q^{(i,j)}_\alpha(kT) + q^{(i,j)}_\beta(kT)}{2} & \text{if both } \alpha \text{ and } \beta \text{ exist,} \\[2mm] \dfrac{q^{(i,j)}_{(i,j)}(kT) - \bar{d} + q^{(i,j)}_\beta(kT)}{2} & \text{if only } \beta \text{ exists,} \\[2mm] \dfrac{q^{(i,j)}_\alpha(kT) + q^{(i,j)}_{(i,j)}(kT) + \bar{d}}{2} & \text{if only } \alpha \text{ exists,} \end{cases} \tag{3.22}$$

where

$$\bar{d} := \begin{cases} d_1 & \text{for } j=2,3,\ldots,n, \\ d_2 & \text{for } j=1. \end{cases}$$

To prevent the situation where a follower settles at the wrong side, i.e., the left-hand side, of its leader, we modify $\mathcal{D}_{(i,j)}(kT)$. First, we pick scalars $\varepsilon > 0$ and $v_0 > 0$ such that $0 < \varepsilon < v_0 T < (r_c - \bar{d})$. Next, if $q^{(i,j)}_{(i,1)}(kT) - q^{(i,j)}_{(i,j)}(kT) \in [0, \bar{d}+\varepsilon]$, we put

$$\mathcal{D}_{(i,j)}(kT) = \frac{q^{(i,j)}_{(i,j)}(kT) + q^{(i,j)}_{(i,1)}(kT) + v_0 T}{2}. \tag{3.23}$$

Based on (3.22), we also introduce $\bar{\mathcal{U}}_{(i,j)}(kT)$ as follows:

$$\text{if } j=1, p_{(1,1)}(kT) \in D_{(i,j),r}(kT), q^{(i,j)}_{(1,1)}(kT) \leq q^{(i,j)}_{(i,j)}(kT),$$
$$\text{then } \bar{\mathcal{U}}_{(i,j)}(kT) = (q^{(i,j)}_{(i,j)}(kT) + q^{(1,1)}_{(1,1)}(kT) + v_0 T))/2;$$
$$\text{if } j \neq 1, p_{(1,1)}(kT) \in D_{(i,j),r}(kT), q^{(i,j)}_{(i,1)}(kT) \leq q^{(i,j)}_{(i,j)}(kT), \tag{3.24}$$
$$\text{then } \bar{\mathcal{U}}_{(i,j)}(kT) = (q^{(i,j)}_{(i,j)}(kT) + q^{(i,j)}_{(i,1)}(kT) + v_0 T))/2;$$
$$\text{otherwise } \bar{\mathcal{U}}_{(i,j)}(kT) = \mathcal{D}_{(i,j)}(kT),$$

where the scalar v_0 is chosen so that $0 < v_0 < \min(r - d_1, r - d_2)/T$.

In order to reach consensus, the coordination variable $\phi_{(i,j)}(\cdot)$ is updated as follows:

$$\phi_{(i,j)}((k+1)T) := \hat{\mathscr{A}}_{(i,j)}(kT). \tag{3.25}$$

At time kT, we define $\hat{\mathscr{F}}_{(i,j)}((k+1)T)$ as the desired value of $\mathscr{F}_{(i,j)}$ for sensor (i,j) at the next time step $(k+1)T$. Similarly, we define $\hat{q}_{(i,j)}((k+1)T)$ as the desired projected (on the line $\mathscr{L}_{(i,j)}(kT)$) location of sensor (i,j) at the next time step $(k+1)T$. These values are chosen to be

$$\begin{aligned}
\hat{\mathscr{F}}_{(i,j)}((k+1)T) &:= \hat{\mathscr{U}}_{(i,j)}(kT), \\
\hat{q}_{(i,j)}((k+1)T) &:= \tilde{\mathscr{U}}_{(i,j)}(kT).
\end{aligned} \tag{3.26}$$

In order to achieve (3.26), we define the velocity components parallel and orthogonal to the line $\mathscr{L}_{(i,j)}$ as

$$\begin{aligned}
\bar{v}_{(i,j)}(kT) &= (\bar{U}_{(i,j)}(kT) - q^{(i,j)}_{(i,j)}(kT))/T, \\
\hat{v}_{(i,j)}(kT) &= (\hat{U}_{(i,j)}(kT) - \mathscr{F}_{(i,j)}(kT))/T
\end{aligned} \tag{3.27}$$

for $i = 1, 2, \ldots, K$, $j = 1, 2, \ldots, n$.

Using (3.27), we introduce the following set of decentralized control laws:

$$\begin{aligned}
v_{(1,1)}(kT) &= 0, \quad \theta_{(1,1)}(kT) = \bar{\phi}; \\
v_{(i,j)}(kT) &= \sqrt{\bar{v}_{(i,j)}(kT)^2 + \hat{v}_{(i,j)}(kT)^2}, \\
\theta_{(i,j)}(kT) &= \\
& \begin{cases} \phi_{(i,j)}(kT) + \xi_{(i,j)}(kT) - \pi/2 & \text{if } \hat{v}_{(i,j)}(kT) \geq 0, \\ \phi_{(i,j)}(kT) - \xi_{(i,j)}(kT) - \pi/2 & \text{if } \hat{v}_{(i,j)}(kT) < 0, \end{cases}
\end{aligned} \tag{3.28}$$

for $i = 1, 2, \ldots, K$ and $j = 1, 2, \ldots, n$, where $\xi_{(i,j)}(kT) := \cos^{-1}(\bar{v}_{(i,j)}(kT)/v_{(i,j)}(kT))$.

3.4 Mathematical Analysis of the K-Barrier Coverage Algorithm

In this section, the main result of this chapter is presented. Before doing so, we first state and prove the following lemma.

Lemma 3.1 *Consider n mobile robotic sensors that belong to group $i \in \{1, 2, \ldots, K\}$, and consider the decentralized control law (3.28). Suppose that Assumption 3.1 holds. If the position of sensor $(i,1)$ satisfies $p_{(i,1)}(kT) \equiv \tilde{p}$ for all $k \geq 0$, then there exists $\tilde{\phi} \in [0, \pi)$ such that*

$$\lim_{k \to \infty} p_{(i,y_j^{(i)})}(kT) = \tilde{p} + jd_1 l(\tilde{\phi} - \pi/2), \qquad j = 1, 2, \ldots, n-1, \tag{3.29}$$

where $\{y_1^{(i)}, y_2^{(i)}, \ldots, y_{n-1}^{(i)}\}$ is some permutation of the set $\{2, 3, \ldots, n\}$.

In other words, for given $i \in \{1, 2, \ldots, K\}$, $\tilde{p}$ and d_1, Lemma 3.1 states that sensors $(i, 2), (i, 3), \ldots, (i, n)$ converge to the line $\{p \in \mathbb{R}^2 : l(\tilde{\phi})^T(p - \tilde{p}) = 0\}$ hosting sensor $(i, 1)$. In addition, the distance between the sensors is d_1.

Proof of Lemma 3.1: The graph's connectivity property imposed in Assumption 3.1 and the update law for $\phi_{(i,j)}(\cdot)$ in (3.25) guarantee that sensors $(i, 1), \ldots, (i, n)$ reach a consensus on the value of the coordination variable $\phi_{(i,j)}(\cdot)$ (see, e.g., [53]): there exists $\tilde{\phi} \in [0, \pi)$ such that

$$\lim_{k \to \infty} \phi_{(i,j)}(kT) = \tilde{\phi}, \quad j = 1, 2, \ldots, n. \tag{3.30}$$

Similarly, since $\phi_{(i,j)}(kT) \to \tilde{\phi}$ and $p_{(i,1)}(kT) \equiv \tilde{p}$, the update rule for $\mathscr{F}_{(i,j)}(\cdot)$ in (3.26) yields that

$$\lim_{k \to \infty} \mathscr{F}_{(i,j)}(kT) = \bar{\mathscr{F}}_i, \quad j = 2, 3, \ldots, n, \tag{3.31}$$

where

$$l(\tilde{\phi})^T p_{(i,1)}(kT) = l(\tilde{\phi})^T \tilde{p} = \bar{\mathscr{F}}_i.$$

In other words, conditions (3.30)–(3.31) guarantee that there exists a line

$$\mathscr{L}_i = \{p \in \mathbb{R}^2 : l(\tilde{\phi})^T p = \bar{\mathscr{F}}_i\} \tag{3.32}$$

such that $p_{(i,1)} \in \mathscr{L}_i$ and

$$\lim_{k \to \infty} d(p_{(i,j)}(kT), \mathscr{L}_i) = 0, \quad j = 2, 3, \ldots, n, \tag{3.33}$$

where $d(p_{(i,j)}(\cdot), \mathscr{L}_i)$ is the distance between the point $p_{(i,j)}(\cdot)$ and the line $\mathscr{L}_i$. Thus, all the sensors (i, j), $j = 2, 3, \ldots, n$, converge to the line $\mathscr{L}_i$ that contains the point $\tilde{p}$.

Next, we are going to show that the distance between the sensors converges to d_1. To this end, we denote by $d_{\max}(kT)$ the largest distance from the line $\mathscr{L}_i$ to $p_{(i,j)}(kT)$'s, i.e., $d_{\max}(kT) := \max_{j=1,2,\ldots,n} d(p_{(i,j)}(kT), \mathscr{L}_i)$. For a given $\delta > 0$, relation (3.33) implies that there exists $\mathscr{J} \geq 0$ such that $d_{\max}(kT) < \delta$ for all $k \geq \mathscr{J}$. Let $q_{(i,j)}(kT)$ be the projection of the position of sensor (i, j) on the line $\mathscr{L}_i$ at time kT; this projection is given by

$$q_{(i,j)}(kT) = l(\tilde{\phi} - \pi/2)^T p_{(i,j)}(kT). \tag{3.34}$$

Thus, (3.26) guarantees that there exists a permutation $\{y_1^{(i)}, y_2^{(i)}, \ldots, y_{n-1}^{(i)}\}$ of the set $\{2, 3, \ldots, n\}$ for which

$$q_0 < q_{(i, y_1^{(i)})}(kT) < q_{(i, y_2^{(i)})}(kT) < \cdots < q_{(i, y_{n-1}^{(i)})}(kT) \tag{3.35}$$

for all $k \geq \mathscr{J}$, where $q_0 := l(\tilde{\phi})^T \tilde{p} = l(\tilde{\phi})^T p_{(i,1)}(kT)$. The condition (3.35) holds for almost all initial positions of the sensors.

Using (3.35) and the fact that $G_i(kT)$ is connected for all $k \geq 0$, we have $q_{(i,y_1^{(i)})}(kT)$
$- q_0 \leq r$ and $q_{(i,y_j^{(i)})}(kT) - q_{(i,y_{j-1}^{(i)})}(kT) \leq r$ for all $k \geq \mathscr{J}$ and $j = 2,3,\ldots,n-1$.
The update rule (3.22) can then be written as a linear dynamic system

$$q((k+1)T) = Aq(kT) + b \quad \text{for } k \geq \mathscr{J}, \tag{3.36}$$

where

$$q(kT) := \left[q_{(i,y_1^{(i)})}(kT) \quad q_{(i,y_2^{(i)})}(kT) \quad \cdots \quad q_{(i,y_{n-1}^{(i)})}(kT) \right]^T,$$

$$b := \left[q_0/2 \quad 0 \quad \cdots \quad d_1/2 \right]^T,$$

and A is a symmetric square matrix with elements $A_{i,j}, 1 \leq i,j \leq n-1$ such that
$A_{i,i+1} = A_{i+1,i} = 1/2$ for $1 \leq i \leq n-2$ and $A_{n-1,n-1} = 1/2$, $A_{i,j} = 0$ for all other
i,j. By using the result [81, p. 514] and the fact that $A_{n-1,n-1} = A_{n-1,n-2} = 1/2$,
it can be shown that $\lim_{k\to\infty} q_{(i,y_j^{(i)})}(kT) = q_0 + jd_1$, $j = 1,2,\ldots,n-1$. Therefore,
$\lim_{k\to\infty} P_{(i,y_j^{(i)})}(kT) = \tilde{p} + jd_1 l(\tilde{\phi} - \pi/2)$ for $j = 1,2,\ldots,n-1$. This completes the
proof of Lemma 3.1. ■

Now we are in position to present the main result.

Theorem 3.1 *Consider Kn mobile robotic sensors described by equation (3.2). Let
Assumptions 3.1–3.2 hold. Then the decentralized control law (3.28) is a K-barrier-
coverage coordinated control for the sensors between two given points L_1 and L_2.*

Proof: We first assume that for the group leaders, i.e., sensors $(1,1),(2,1),\ldots,(K,1)$,
the following conditions hold:

$$\phi_{(i,1)}(kT) \equiv \tilde{\phi}, \quad i = 1,2,\ldots,K \tag{3.37}$$

and

$$P_{(1,1)}(kT) \equiv h_{1,1}, \quad P_{(x_i,1)}(kT) \equiv h_{i,1} \tag{3.38}$$

for $k \geq 0$ and $i = 2,3,\ldots,K$, where $\{x_2,x_3,\ldots,x_K\}$ is a permutation of the set $\{2,3,\ldots,K\}$. By direct application of Lemma 3.1, it is then straightforward to show that

$$\lim_{k\to\infty} P_{(1,y_j^{(1)})}(kT) = h_{1,1} + (j-1)d_1 l(\tilde{\phi} - \pi/2) = h_{1,j},$$

$$\lim_{k\to\infty} P_{(x_i,y_j^{(x_i)})}(kT) = h_{i,1} + (j-1)d_1 l(\tilde{\phi} - \pi/2) = h_{i,j} \tag{3.39}$$

for $i = 2,3,\ldots,K$ and $j = 2,3,\ldots,n$, where

$$\{y_1^{(1)}, y_2^{(1)}, \ldots, y_{n-1}^{(1)}\} \quad \text{and} \quad \{y_1^{(x_i)}, y_2^{(x_i)}, \ldots, y_{n-1}^{(x_i)}\}$$

are permutations of the set $\{2,3,\ldots,n\}$.

To complete the proof, we just need to show that the conditions (3.37)–(3.38) do
hold as $k \to \infty$. In fact, this can be done by setting $p_{(1,1)} = L_1 = h_{1,1}$ and $\phi_{(1,1)}(kT) =$

$\bar{\phi}$ for all $k \geq 0$ and by following the steps in the proof of Lemma 3.1. Then it can be shown that, for $i = 1, 2, \ldots, K$,

$$\lim_{k \to \infty} p_{(x_i,1)}(kT) = h_{1,1} + id_2 l(\bar{\phi} - \pi) = h_{i+1,1}, \tag{3.40}$$

where $x_1 = 1$. Finally, equations (3.39), (3.40) mean that the requirements of Definition 3.1 are fulfilled. This completes the proof of Theorem 3.1. ■

In this chapter, our focus is on providing a mathematical development of a decentralized control for a group of robotic sensors to achieve K-barrier coverage. However, there are some implementation issues that need to be considered. For example, sensors may not have access to perfect locations or coordinates of themselves and neighbors due to noisy or limited measurements. In this regard, a number of existing techniques, e.g., coordinates estimation from range-only measurements [86, 87], can be adopted. In relation to the issue of estimation using limited measurements, there is a number of techniques in the literature that can also be employed; see e.g., [73, 104]. Furthermore, methods of range-only based robot navigation can be applied [77, 116]. Another practical issue is the drive mechanism of the mobile sensors. In this chapter, we assume that the sensors have holonomic drive mechanism (i.e., they can move equally well in any direction). At the same time, there are robotic sensors that are non-holonomic. However, our holonomic drive mechanism assumption is not uncommon for ground vehicles; see e.g., [57]. For controlling a group of non-holonomic robots, a comprehensive treatment can be found in [30].

3.5 Illustrative Examples

In this section, we present several simulation results for various scenarios to illustrate the effectiveness of the proposed algorithm. In the first simulation, our objective is to obtain K-barrier coverage with $K = 4$ and $n = 6$. The system parameters for simulation are given by $r = 1.5$, $d_1 = 1$, $d_2 = 0.8$, and $\bar{\phi} = \pi/3$. Figure 3.3(a) shows the positions of the sensors at the initial deployment ($\lozenge$) at $k = 0$ and at final deployment ($\circ$) at $k = 200$. Also, Fig. 3.3(a) shows the sensors' trajectories. The velocities of the sensors are shown in Fig. 3.3(b). Initially, the sensors were randomly deployed near L_1 except for sensor $(1, 1)$, which was dispatched at L_1. By applying our algorithm, they formed K layers of sensor barriers between L_1 and L_2 that satisfy (3.10), (3.11). Similar results were obtained with different K, n, and $\bar{\phi}$; they are shown in Figs. 3.4 and 3.5. Again, the sensors formed K layers of sensor barriers in each scenario. In other words, any intruder's path that is orthogonal to the line segment between L_1 and L_2 is detected by at least K sensors. In fact, any intruder's path that is within the region

$$\mathscr{C} := \left\{ p \in \mathbb{R}^2 : l(\beta)^T L_1 \leq l(\beta)^T p \leq l(\beta)^T L_2 \right\}, \tag{3.41}$$

where $\beta = \bar{\phi} - \pi/2$, is also detected by at least K sensors.

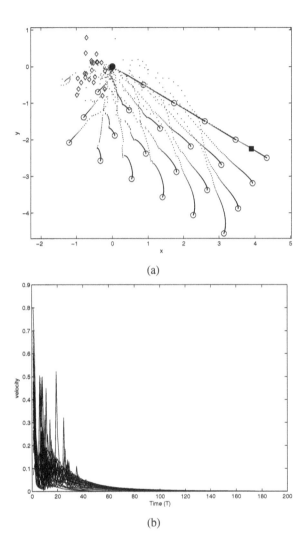

(a)

(b)

Figure 3.3 K-barrier coverage between two points $L_1(\bullet)$ and $L_2(\blacksquare)$ with $K = 4$, $n = 6$, $r = 1.5$, $d_1 = 1$, $d_2 = 0.8$, and $\bar{\phi} = \pi/3$: (a) Initial deployment ($\Diamond$) at $k = 0$ and final deployment ($\circ$) at $k = 200$; (b) The velocities of the mobile sensors.

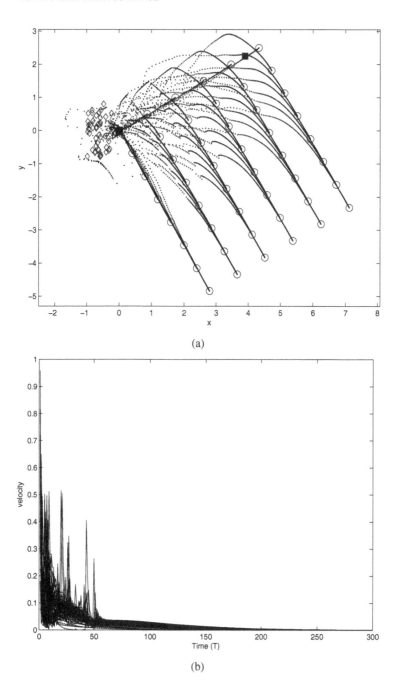

Figure 3.4 K-barrier coverage between two points $L_1(\bullet)$ and $L_2(\blacksquare)$ with $K = 8$, $n = 6$, $r = 1.5$, $d_1 = 1$, $d_2 = 0.8$, and $\bar{\phi} = 2\pi/3$: (a) Initial deployment ($\Diamond$) at $k = 0$ and final deployment ($\circ$) at $k = 300$; (b) The velocities of the sensors.

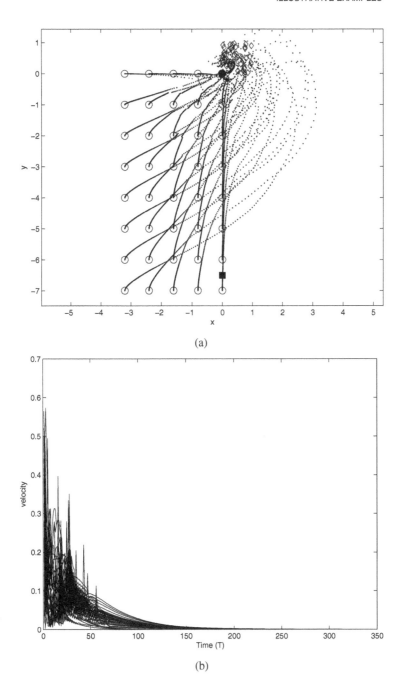

(a)

(b)

Figure 3.5 *K*-barrier coverage between two points L_1 (●) and L_2 (■) with $K = 5$, $n = 8$, $r = 2$, $d_1 = 1$, $d_2 = 0.8$, and $\bar{\phi} = 0$: (a) Initial deployment (◊) at $k = 0$ and final deployment (○) at $k = 350$; (b) The velocities of the sensors.

CHAPTER 4

PROBLEMS OF BARRIER AND SWEEP COVERAGE IN CORRIDOR ENVIRONMENTS

4.1 Introduction

In this chapter, we focus on the problems of barrier coverage and sweep coverage in a corridor environment using a network of mobile robotic sensors. As indicated in [35], these two coverages have an enormous potential for many applications. They include, but are not limited to, a military; for example, barrier coverage can be applied to mine deployment and sentry duty. Meanwhile, sweep coverage can be used in multi-sensor minesweeping [1, 12, 36], reconnaissance, maintenance inspection, and ship hull cleaning. Our barrier coverage problem is to deploy a group of mobile autonomous robotic sensors to form a barrier that detects intruders trying to enter a protected region in a corridor. On the other hand, the sweep coverage problem is to deploy the mobile robotic sensors so that every point in a specific region in a corridor is detected by the sensors as they move along the region.

We consider the mobile robotic sensors that are self-deployed, as the optimal placement of robotic sensors may not be achieved at the initial deployment in the barrier coverage and also the robotic sensors are required to maintain a certain formation in sweep coverage to facilitate detection. In addition, each mobile robotic sensor may have severe detection constraints and may sense only a small portion

Decentralized Coverage Control Problems for Mobile Robotic Sensor and Actuator Networks. **33**
By Andrey V. Savkin, Teddy M. Cheng, Zhiyu Xi, Faizan Javed, Alexey S. Matveev, and Hung Nguyen. Copyright © 2015 by the Institute of Electrical and Electronics Engineers, Inc.

of the environment. Hence, we are particularly interested in self-deployed mobile robotic sensors that work in a distributed and unsupervised mode.

The objective of this chapter is to develop a decentralized control law for a group of self-deployed mobile robotic sensors to perform barrier coverage and sweep coverage tasks in a corridor environment. For barrier coverage, our proposed control laws drive the robotic sensors to form a sensor barrier across a given corridor from any initial deployment. In addition, the robotic sensors will be evenly distributed in the sensor barrier. As for sweep coverage, the sensor barrier will move along the corridor at a given speed while maintaining the line formation so that any object that lies in the path of the moving sensor barrier is detected; this type of sweep coverage is particularly useful in mine sweeping.

To achieve barrier coverage in a rectangular environment, control of mobile sensors using a virtual force model was studied by Shen *et al.* [112], but only simulation results were presented and theoretical validation of the proposed algorithm was absent. In this chapter, we propose a decentralized cooperative control law for motion coordination of mobile robotic sensors for both barrier coverage and sweep coverage in a corridor environment and, most importantly, we provide a detailed theoretical development of the control law using the information consensus approach [53]. Cortés *et al.* [29] studied a problem of coverage control to achieve optimal sensor placement of mobile sensors. The problem can be considered as a blanket coverage problem, as defined by Gage [35], and belongs to the area of locational optimization. To solve it, a desired sensor distribution in the sensing environment is required to be known by all sensors a priori. However, the robotic sensors considered in this chapter may not have such information a priori.

The main results of the chapter are originally published in [21].[1]

The rest of the chapter is organized as follows. In Section 4.2, we formulate the problem of decentralized control of mobile robotic sensors for barrier coverage and sweep coverage in corridor environments. Section 4.3 addresses a barrier coverage problem in a one-dimensional space; its findings are useful for the development of the main results. Furthermore, we study a two-dimensional barrier coverage problem in Section 4.4. Then, the results on barrier coverage are extended to sweep coverage and presented in Section 4.5. Section 4.6 offers some computer simulation results to demonstrate the effectiveness of the proposed algorithms.

4.2 Corridor Coverage Problems

Our robotic sensor deployment problem is to develop a distributed motion coordinated algorithm for coverage in a region with autonomous mobile robotic sensors. We consider a two-dimensional region $\mathscr{C} \subset \mathbb{R}^2$ between two parallel lines W_1 and W_2, referred to as a *corridor*. Let $l = [l_1 \; l_2]^T$ be a given vector, and let $d_1 > d_2$ be

---

[1]Cheng T.M. and Savkin A.V.: Decentralized control for mobile robotic sensor network self-deployment: Barrier and sweep coverage problems. Robotica. **29(2)**, 283-294 (March 2011). Copyright ©2010 Cambridge University Press. Reprinted with permission.

given scalars associated with the lines W_1 and W_2, respectively. The corridor $\mathscr{C}$ is defined as the intersection of the two regions defined by the lines W_1 and W_2:

$$
\mathscr{C} := \left\{ (x,y)^T \in \mathbb{R}^2 : l^T \begin{bmatrix} x \\ y \end{bmatrix} \leq d_1 \right\} \cap \left\{ (x,y)^T \in \mathbb{R}^2 : l^T \begin{bmatrix} x \\ y \end{bmatrix} \geq d_2 \right\}. \qquad (4.1)
$$

Let θ_0 be the angle of the corridor $\mathscr{C}$ with respect to the x-axis. Without any loss of generality, we assume that $\theta_0 \in [0, \pi)$.

4.2.1 Barrier Coverage

The first self-deployment objective is to build a network of autonomous mobile robotic sensors to cover a segment B across the corridor $\mathscr{C}$. This kind of coverage forms a robotic sensor barrier that can be used to detect objects (intruders) penetrating in the corridor $\mathscr{C}$. At initial deployment, the sensors scatter over $\mathscr{C}$ and may not detect intruders moving along $\mathscr{C}$; see Fig 4.1(a). To meet the self-deployment objective, the mobile sensors must move autonomously to cover a line segment B so that the robotic sensor network guarantees detection of any intruder crossing the segment B; see Fig 4.1(b). According to the terminology introduced by Gage [35], this deployment problem is called a barrier coverage problem.

We consider a mobile sensor network consisting of n autonomous robotic sensors, labelled 1 through n. To avoid repetition, we shall call the mobile robotic sensors simply as sensors. Let $v_i(t)$ and $\theta_i(t)$ be the linear speed and heading of sensor i, respectively. The kinematic equations of the sensors are given by

$$
\begin{aligned}
\dot{x}_i(t) &= v_i(t)\cos(\theta_i(t)), \\
\dot{y}_i(t) &= v_i(t)\sin(\theta_i(t))
\end{aligned} \qquad (4.2)
$$

for $i = 1, 2, \ldots, n$, where $x_i(\cdot)$ and $y_i(\cdot)$ are the Cartesian coordinates of sensor i, and $\theta_i(\cdot) \in \mathbb{R}$ is its heading measured from the x-axis in the counterclockwise direction. The speed v_i and heading θ_i satisfy $|v_i(t)| \leq v_{\max}$ and $\theta_i(t) \in [0, \pi)$.

Given a sampling period $T > 0$, the discrete-time dynamics of sensor i are described by the following model:

$$
\begin{aligned}
x_i((k+1)T) &= x_i(kT) + \int_{kT}^{(k+1)T} v_i(t)\cos(\theta_i(t))\, dt, \\
y_i((k+1)T) &= y_i(kT) + \int_{kT}^{(k+1)T} v_i(t)\sin(\theta_i(t))\, dt,
\end{aligned} \qquad (4.3)
$$

where the heading $\theta_i(\cdot)$ and speed $v_i(\cdot)$ are the control inputs. At time kT, each sensor i detects other sensors in a range of $r > 0$, i.e., the sensors from the disk

$$
D_{i,r}(kT) := \{ (x,y)^T \in \mathbb{R}^2 : (x - x_i(kT))^2 + (y - y_i(kT))^2 \leq r^2 \}.
$$

Let $\mathcal{N}_i(kT)$ be the set of all such sensors $j \neq i$, and let $|\mathcal{N}_i(kT)|$ be the number of them. These elements are called *neighbors* of sensor i at time kT. Any sensor can also detect the boundary $\partial \mathscr{C}$ of the corridor $\mathscr{C}$ within a range of R.

Assumption 4.1 *The sensing range r and the boundary sensing range R satisfy*

$$r < v_{\max}T/2 \quad and \quad R > r\sqrt{2}. \tag{4.4}$$

For given $v_{\max}$ and r, the condition $r < v_{\max}T/2$ in (4.4) can be met by choosing an appropriate sampling period T.

Assumption 4.2 *The initial heading $\theta_i(0)$ of each sensor $i = 1, 2, \dots, n$ and the angle of the corridor orientation θ_0 satisfy the following condition:*

$$\cos(\theta_0 - \theta_i(0)) > 0. \tag{4.5}$$

Assumption 4.3 *Let $d > 0$ stand for the width of the corridor. The number n of the sensors and the radius r of the sensing region satisfy*

$$(n+1)r > d / \min_{i=1,2,\dots,n} \left\{ \cos(\theta_0 - \theta_i(0)) \right\}. \tag{4.6}$$

Let $\mathscr{P}$ be the collection of all simple undirected graphs defined on n vertices. For any time $kT \geq 0$, the relationships between neighbors are described by a simple undirected graph $G(kT) \in \mathscr{P}$ with vertex set $\{1, 2, \dots, n\}$, where i corresponds to sensor i. The vertices $i \neq j$ of the graph are connected by an edge if and only if the sensors i and j are neighbors at time kT.

We suppose that the graph sequence $G(kT)$ satisfies the standard Main Connectivity Assumption (i.e., Assumption 2.4 from Chapter 2).

To be cost effective, each mobile sensor may have a very limited communication capability. So the control laws for the mobile sensors should be distributed or decentralized in the sense that the movement of each sensor only relies on information about its neighbors and itself. We assume that the information available to sensor i at time kT is $\{x_j(kT), y_j(kT), \theta_j(kT)\}$, where j ranges over $\mathscr{N}_i(kT) \cup \{i\}$. Our first aim is to design coordinated control laws that steer the sensors so that they become eventually distributed along a segment B crossing the corridor $\mathscr{C}$.

To specify this aim, we introduce the vector

$$l = [\sin(\theta_0) \quad -\cos(\theta_0)]^T \tag{4.7}$$

and the line

$$\mathscr{L}_0 := \{(x,y)^T \in \mathbb{R}^2 : x\cos(\theta_0) + y\sin(\theta_0) = \mathscr{F}_0\} \tag{4.8}$$

perpendicular to the corridor $\mathscr{C}$ with the orientation angle θ_0. Here $\mathscr{F}_0$ is some scalar. Next, we define n points $h_i \in \mathbb{R}^2$ on $\mathscr{L}_0$ by (see Fig. 4.6(a))

$$h_i = h_0 + i \left(\frac{d_1 - d_2}{n+1} \right) l, \quad i = 1, 2, \dots, n, \tag{4.9}$$

where

$$h_0 := \mathscr{L}_0 \cap \left\{ (x,y)^T \in \mathbb{R}^2 : \sin(\theta_0)x - \cos(\theta_0)y = d_2 \right\}.$$

Let $p_i(kT) = (x_i(kT), y_i(kT))^T$ be the position of sensor i at time kT.

Definition 4.1 (Barrier Coverage) *Given a corridor $\mathscr{C}$ with an orientation angle θ_0, a decentralized control law is called an* optimal corridor-barrier-coverage coordinated control *for the mobile sensors if for almost all initial sensor positions, there exists a permutation of the set $\{1, 2, \ldots, n\}$, denoted by $\{z_1, z_2, \ldots, z_n\}$, such that the following condition holds:*

$$\lim_{k \to \infty} \| p_{z_i}(kT) - h_i \| = 0, \quad i = 1, 2, \ldots, n. \tag{4.10}$$

Using the definition introduced by Kumar *et al.* [59], a given corridor is k_0-*barrier covered* under the optimal corridor-barrier-coverage coordinated control, where $k_0 := \lfloor (n+1)r/d \rfloor$ and $\lfloor x \rfloor$ is the largest integer that is less than or equal to x.

4.2.2 Sweep Coverage

The second self-deployment problem addressed in this chapter is to deploy a network of mobile sensors so that they move in a formation and detect every point in a specified region. Again, the sensors are randomly scattered over a corridor $\mathscr{C}$ at initial deployment. One way to achieve the deployment objective is to autonomously move the sensors in a line formation and to scan the specified region as they go along. This second coverage problem is of sweep coverage type, as was defined by Gage [35].

This sweep coverage problem extends the corridor barrier coverage described in the previous section, except that the sensors are required to move along a corridor at a desired speed v_0. We assume that this speed is known to all the sensors.

Assumption 4.4 *The sweeping speed v_0 satisfies*

$$|v_0| \leq \min\left\{ \left(\frac{v_{\max} T}{2} - r \right), \, (\sqrt{R^2 - r^2} - r) \right\} / T. \tag{4.11}$$

In sweep coverage, the sensors converge to a line that is not stationary; which is in contrast to the barrier coverage; cf. (4.8). Therefore, we define the following line $\mathscr{L}_0(kT)$, which moves as k ranges over $0, 1, \ldots$:

$$\mathscr{L}_0(kT) := \left\{ (x, y)^T \in \mathbb{R}^2 : x\cos(\theta_0) + y\sin(\theta_0) = \mathscr{F}_0 + kTv_0 \right\}, \tag{4.12}$$

where $\mathscr{F}_0$ is some scalar. We also define n points $h_i(kT)$ on this line (see Fig. 4.6(b)) by

$$h_i(kT) = h_0(kT) + i\left(\frac{d_1 - d_2}{n+1} \right)l, \quad i = 1, 2, \ldots, n, \tag{4.13}$$

where

$$h_0(kT) := \mathscr{L}_0(kT) \cap \left\{ (x, y)^T \in \mathbb{R}^2 : \sin(\theta_0)x - \cos(\theta_0)y = d_2 \right\}.$$

Definition 4.2 (Sweep Coverage) *Given a corridor $\mathscr{C}$ with angle θ_0 and a desired sweeping speed v_0, a decentralized control law is called an* optimal corridor-sweep-coverage coordinated control with sweeping speed v_0 *for the mobile sensors if for almost all initial sensor positions, there exists a permutation of the set $\{1, 2, \ldots, n\}$, denoted by $\{z_1, z_2, \ldots, z_n\}$, such that the following condition holds:*

$$\lim_{k \to \infty} \| p_{z_i}(kT) - h_i(kT) \| = 0, \quad i = 1, 2, \ldots, n. \tag{4.14}$$

4.3 Barrier Coverage in 1D Space

In this section, we propose a decentralized control law to deploy mobile sensors in a one-dimensional space so that all inter-neighbor distances are equalized. Without any loss of generality, we assume that the n number of mobile sensors have been already aligned on the x-axis between two fixed points x_0 and x_{n+1}; their initial positions are assumed to be

$$x_0 < x_1(0) < x_2(0) < \ldots < x_n(0) < x_{n+1}. \tag{4.15}$$

For any sensor i, we introduce the following update rule:

if $x_i(kT) - x_{i-1}(kT) \le r$ and $x_{i+1}(kT) - x_i(kT) \le r$,
  then $x_i((k+1)T) = \dfrac{x_{i-1}(kT) + x_{i+1}(kT)}{2}$;
if $x_{i+1}(kT) - x_i(kT) \le r$ and $x_i(kT) - x_{i-1}(kT) > r$,
  then $x_i((k+1)T) = \dfrac{x_i(kT) - r + x_{i+1}(kT)}{2}$;   (4.16)
if $x_i(kT) - x_{i-1}(kT) \le r$ and $x_{i+1}(kT) - x_i(kT) > r$,
  then $x_i((k+1)T) = \dfrac{x_{i-1}(kT) + x_i(kT) + r}{2}$;
if $x_i(kT) - x_{i-1}(kT) > r$ and $x_{i+1}(kT) - x_i(kT) > r$,
  then $x_i((k+1)T) = x_i(kT)$.

Here $k = 0, 1, 2, \ldots$ and $x_0(kT) \equiv x_0$, $x_{n+1}(kT) \equiv x_{n+1}$.

Lemma 4.1 *Consider n mobile sensors with sensing range $r > 0$. Suppose that their initial locations satisfy the condition (4.15) and also that n and r satisfy $(n+1)r > L := x_{n+1} - x_0$. Then the update rule (4.16) ensures that*

$$\lim_{k \to \infty} x_i(kT) = x_0 + \frac{iL}{n+1} \quad for \quad i = 1, 2, \ldots, n. \tag{4.17}$$

Proof: First we prove that there exists an integer k_0 such that the condition

$$x_{i+1}(kT) - x_i(kT) < r \tag{4.18}$$

holds for $i = 1, 2, \ldots, n - 1$ and all $k \geq k_0$. Indeed, it immediately follows from the rule (4.16) that if (4.18) holds for some $k = k_0$, then it also holds for all $k > k_0$. Furthermore, condition $L < (n+1)r$ and the rule (4.16) imply that (4.18) does hold for some $k = k_0$. Now if (4.18) holds for some k, then the rule (4.16) is obviously reduced to the linear relationship

$$x((k+1)T) = Ax(kT) + b, \tag{4.19}$$

where

$$x(kT) := \begin{bmatrix} x_1(kT) \\ x_2(kT) \\ \vdots \\ x_{n-1}(kT) \\ x_n(kT) \end{bmatrix}, \qquad b := \begin{bmatrix} x_0/2 \\ 0 \\ \vdots \\ 0 \\ x_{n+1}/2 \end{bmatrix},$$

and A is a symmetric $n \times n$ matrix with elements a_{ij} such that $a_{i,i+1} = a_{i+1,i} = \frac{1}{2}$ for all $1 \leq i \leq n - 1$, and $a_{ij} = 0$ for all other i, j. Since the matrix A is symmetric, all its eigenvalues λ are real. Furthermore, it follows from the Perron–Frobenius theorem [49] that $\max |\lambda| \leq \frac{1}{2}$. Hence the matrix A is stable, the system (4.19) has a unique equilibrium point $\tilde{x}$, and all the solutions of the system converge to this point. Meanwhile, it is easy to see that this equilibrium point is $\tilde{x} = [\tilde{x}_1 \ \tilde{x}_2 \ldots \tilde{x}_n]^T$, where $\tilde{x}_i = x_0 + iL/(n+1)$. This completes the proof of the lemma. ∎

4.4   Corridor Barrier Coverage

At any time $t = kT$, $k = 0, 1, 2, \ldots$, the following operations are performed.

We first define the average heading of sensor i and its neighbors as

$$\mathscr{A}_i(kT) := \frac{1}{1 + |\mathscr{N}_i(kT)|} \left(\theta_i(kT) + \sum_{j \in \mathscr{N}_i(kT)} \theta_j(kT) \right) \tag{4.20}$$

and introduce the variable

$$\mathscr{H}_i(kT) = \begin{cases} \mathscr{A}_i(kT) & \text{if } \partial\mathscr{C} \cap D_{i,R}(kT) = \emptyset, \\ \theta_0 & \text{if } \partial\mathscr{C} \cap D_{i,R}(kT) \neq \emptyset, \end{cases} \tag{4.21}$$

where $\partial\mathscr{C}$ stands for the boundary of the corridor $\mathscr{C}$. For each neighbor $j \in \mathscr{N}_i(kT)$ of sensor i, we define the following scalar:

$$c_{i,j}(kT) = \begin{bmatrix} \cos(\mathscr{H}_i(kT)) & \sin(\mathscr{H}_i(kT)) \end{bmatrix} \begin{bmatrix} x_j(kT) \\ y_j(kT) \end{bmatrix}. \tag{4.22}$$

Similarly to the average heading (4.20), we define the average of the scalars $c_{i,j}(\cdot)$:

$$\mathcal{M}_i(kT) := \frac{1}{1+|\mathcal{N}_i(kT)|}\left(c_{i,i}(kT) + \sum_{j\in\mathcal{N}_i(kT)} c_{i,j}(kT)\right). \qquad (4.23)$$

Also, we define a scalar $\mathcal{F}_i(kT)$ by

$$\mathcal{F}_i(kT) = \begin{bmatrix}\cos(\mathcal{H}_i(kT)) & \sin(\mathcal{H}_i(kT))\end{bmatrix}\begin{bmatrix}x_i(kT)\\y_i(kT)\end{bmatrix}. \qquad (4.24)$$

If sensor i detects the boundary of the corridor, i.e., $\partial\mathcal{C}\cap D_{i,R}(kT) \neq \emptyset$, the heading $\theta_i(\kappa)$ of this sensor is set to θ_0 for $\kappa = k+1, k+2, \ldots$.

Next, we define a line $\mathcal{L}_i(kT)$ as follows:

$$\mathcal{L}_i(kT) = \{(x,y)^T \in \mathbb{R}^2 : x\cos(\mathcal{H}_i(kT)) + y\sin(\mathcal{H}_i(kT)) = \mathcal{M}_i(kT)\}. \qquad (4.25)$$

We also determine the projection $q^i_j(kT)$ of the position of every sensor $j \in \mathcal{N}_i(kT) \cup \{i\}$ on the line $\mathcal{L}_i(kT)$:

$$q^i_j(kT) = \begin{bmatrix}\sin(\mathcal{H}_i(kT)) & -\cos(\mathcal{H}_i(kT))\end{bmatrix}\begin{bmatrix}x_j(kT)\\y_j(kT)\end{bmatrix}. \qquad (4.26)$$

Then we define $\alpha, \beta \in \mathcal{N}_i(kT)$ such that $q^i_\alpha(kT)$ and $q^i_\beta(kT)$ are immediately next to $q^i_i(kT)$ and $q^i_\alpha(kT) < q^i_i(kT) < q^i_\beta(kT)$. Here either α or β or both may not exist since the set $\mathcal{N}_i(kT)$ may contain only one sensor $|\mathcal{N}_i(kT)| = 1$ or even be empty. If α does not exist, but $\partial\mathcal{C}\cap D_{i,r}(kT) \neq \emptyset$ and one of the intersection points $\mathcal{L}_i(kT)\cap\partial\mathcal{C}$ has the projection less than $q^i_i(kT)$, then the intersection point is taken as α. Similarly, if β does not exist, but $\partial\mathcal{C}\cap D_{i,r}(kT) \neq \emptyset$ and one of the intersection points $\mathcal{L}_i(kT)\cap\partial\mathcal{C}$ has the projection greater than $q^i_i(kT)$, then the intersection point is taken as β.

Using the coordinates $q^i_j(kT)$, we introduce the following quantity $\mathcal{Q}_i(kT)$:

$$\mathcal{Q}_i(kT) = \begin{cases} \dfrac{q^i_\alpha(kT) + q^i_\beta(kT)}{2} & \text{if } \alpha \text{ and } \beta \text{ exist,} \\[2mm] \dfrac{q^i_i(kT) - r + q^i_\beta(kT)}{2} & \text{if only } \beta \text{ exists,} \\[2mm] \dfrac{q^i_\alpha(kT) + q^i_i(kT) + r}{2} & \text{if only } \alpha \text{ exists,} \\[2mm] q^i_i(kT) & \text{if both } \alpha \text{ and } \beta \text{ do not exist.} \end{cases} \qquad (4.27)$$

Now we are in a position to introduce a control law. For $t \in (kT, (k+1/2)T]$, the control inputs $\theta_i(\cdot)$ and $v_i(\cdot)$ are chosen as

$$\theta_i(t) = \begin{cases} \mathcal{H}_i(kT) + \pi/2 & \text{if } \mathcal{H}_i(kT) < \pi/2, \\ \mathcal{H}_i(kT) - \pi/2 & \text{if } \mathcal{H}_i(kT) \geq \pi/2, \end{cases}$$

$$v_i(t) = \begin{cases} -2(\mathcal{Q}_i(kT) - q^i_i(kT))/T & \text{if } \mathcal{H}_i(kT) < \pi/2, \\ 2(\mathcal{Q}_i(kT) - q^i_i(kT))/T & \text{if } \mathcal{H}_i(kT) \geq \pi/2. \end{cases} \qquad (4.28)$$

On the other hand, for $t \in ((k+1/2)T, (k+1)T]$, we choose the heading and velocity control inputs $\theta_i(t)$ and $v_i(t)$ as follows

$$
\begin{aligned}
\theta_i(t) &= \mathcal{H}_i(kT), \\
v_i(t) &= 2\Big(\mathcal{M}_i(kT) - \mathcal{F}_i(kT) \Big)/T,
\end{aligned}
\tag{4.29}
$$

where $\mathcal{H}_i(-T) := \theta_i(0)$.

Intuitively, the control law (4.28) drives the sensor i parallel to the line (4.25) during the time interval $(kT, (k+1/2)T]$ for equalizing the projected (onto (4.25)) distances between adjacent sensors, whereas the control law (4.29) steers the sensor i to the line (4.25) during the time interval $((k+1/2)T, (k+1)T]$. Since the control law (4.28) only drives sensor i parallel to the line (4.25) over the time interval $(kT, (k+1/2)T]$, (4.29) and (4.28) imply that the heading and scalar $\mathcal{F}_i$ at time $(k+1)T$ are as follows:

$$
\begin{aligned}
\theta_i((k+1)T) &= \mathcal{H}_i(kT), \\
\mathcal{F}_i((k+1)T) &= \mathcal{M}_i(kT).
\end{aligned}
\tag{4.30}
$$

Now we are in position to present the first main result of this chapter.

Theorem 4.1 *Consider n mobile sensors described by equation (4.3). Suppose that Assumptions 4.1–4.3 are satisfied. Then the decentralized control law (4.28) and (4.29) is an optimal corridor-barrier-coverage coordinated control for the mobile sensors.*

Proof: We first show that if there exist sensors that detect the boundary of the corridor at some finite time $\rho T \geq 0$, corridor coverage will be achieved. Let $\mathcal{Y}$ be the set of the respective sensors ξ, i.e., such that $\partial \mathcal{C} \cap D_{\xi,R}(\rho T) \neq \emptyset$. Due to the heading control law in (4.29), $\theta_\xi(kT) \equiv \theta_0$ for $\xi \in \mathcal{Y}$ and $k = \rho+1, \rho+2, \ldots$. The sensors from $\mathcal{Y}$ can be viewed as leaders in seeking a consensus on the heading. So by [53, Theorem 4], Assumption 2.4 guarantees that $\lim_{k \to \infty} \theta_i(kT) = \theta_0$ for all $i = 1, 2, \ldots, n$. As for the scalar $\mathcal{F}_i(kT)$, the control law for $v_i(\cdot)$ from (4.29) and [53, Theorem 3] imply that there exists a scalar $\mathcal{F}_0$ such that $\lim_{k \to \infty} \mathcal{F}_i(kT) = \mathcal{F}_0$ for $i = 1, 2, \ldots, n$. So for any $\varepsilon_1, \varepsilon_2 > 0$, there exists $\mathcal{K} \geq \rho$ such that $|\theta_i(kT) - \theta_0| < \varepsilon_1$ and $|\mathcal{F}_i(kT) - \mathcal{F}_0| < \varepsilon_2$ for all $k \geq \mathcal{K}$, $i = 1, 2, \ldots, n$. After introducing the line

$$
\mathcal{L}_0 = \{(x,y)^T \in \mathbb{R}^2 : x\cos(\theta_0) + y\sin(\theta_0) = \mathcal{F}_0\},
\tag{4.31}
$$

we have

$$
\lim_{k \to \infty} d(p_i(kT), \mathcal{L}_0) = 0, \quad i = 1, 2, \ldots, n,
\tag{4.32}
$$

where $p_i(\cdot)$ is the location of sensor i and $d(p_i(\cdot), \mathcal{L}_0)$ is the distance between the point $p_i(\cdot)$ and the line $\mathcal{L}_0$. Since θ_0 is the angle of the corridor $\mathcal{C}$, the line $\mathcal{L}_0$ is orthogonal to $\mathcal{C}$.

Let $d_{\max}(kT)$ be the largest distance from the line $\mathcal{L}_0$ to the sensors: $d_{\max}(kT) := \max_{i=1,2,\ldots,n} d(p_i(kT), \mathcal{L}_0)$. For any $\delta > 0$, there exists $\mathcal{J} \geq \rho$ such that $d_{\max}(kT) <$

δ for all $k \geq \mathscr{J}$, and there also exists a permutation $\{z_1,\ldots,z_n\}$ of the set $\{1,\ldots,n\}$ such that the projections of the positions of sensors $z_1, z_2, \ldots, z_n$ on the line $\mathscr{L}_0$ satisfy the following condition

$$q_0 < q_{z_1}(kT) < q_{z_2}(kT) < \ldots < q_{z_n}(kT) < q_{n+1} \qquad (4.33)$$

for all $k \geq \mathscr{J}$, where $\{q_0, q_{n+1}\} := \mathscr{C} \cap \mathscr{L}_0$. The condition (4.33) holds for almost all initial conditions.

With regard to (4.28), Lemma 4.1 yields that

$$\lim_{k \to \infty} q_{z_i}(kT) = q_0 + \frac{id}{n+1} \qquad (4.34)$$

for $i = 1, 2, \ldots, n$, where $d := q_{n+1} - q_0$. It is clear that $q_0 = d_2$ and $q_{n+1} = d_1$. By (4.9), $l^T h_i = q_0 + id/(n+1)$, and equation (4.34) implies that

$$\lim_{k \to \infty} |q_{z_i}(kT) - l^T h_i| = 0, \quad i = 1, 2, \ldots, n. \qquad (4.35)$$

Thanks to (4.32) and (4.35), we have

$$\lim_{k \to \infty} \|p_{z_i}(kT) - h_i\| = 0, \quad i = 1, 2, \ldots, n \qquad (4.36)$$

since $\|p_{z_i}(kT) - h_i\| \leq d(p_i(kT), \mathscr{L}_0) + |q_{z_i}(kT) - l^T h_i|$. In other words, all the sensors converge to the line $\mathscr{L}_0$, and for any i, there is a sensor z_i converging to the point h_i. By Definition 4.1, the decentralized control law (4.29) and (4.28) is an optimal corridor-barrier-coverage coordinated control.

To complete the proof, it remains to show that there exists a finite time ρT for which $\partial \mathscr{C} \cap D_{i,R}(\rho T) \neq \emptyset$. By invoking Assumption 2.4 once more and using a similar argument as in the foregoing proof, we conclude that the control laws (4.29), (4.28) make the sensors converging to a line $\mathscr{L}$ and equispaced for almost all initial data $(x_i(0), y_i(0), \theta_i(0)), i = 1, 2, \ldots, n$. By Assumption 4.3, such a line satisfies the condition $\mathscr{L} \cap \partial \mathscr{C} \neq \emptyset$, and hence there exists a sensor i and some finite time $\rho T \geq 0$ such that $\partial \mathscr{C} \cap D_{i,R}(\rho T) \neq \emptyset$. This completes the proof of Theorem 4.1. ∎

4.5 Corridor Sweep Coverage

To solve our corridor sweep coverage problem, we de⬚ne the control $v_i(\cdot)$, $i = 1, 2, \ldots, n$, in (4.3) as follows:

$$v_i(t) = \begin{cases} \bar{v}_i(t) & \text{for } t \in (kT, (k+1/2)T], \\ \bar{v}_i(t) + 2v_0 & \text{for } t \in ((k+1/2)T, (k+1)T], \end{cases} \qquad (4.37)$$

where v_0 is the desired sweeping speed and $\bar{v}_i$ is a control input to the following system:

$$\begin{bmatrix} \bar{x}_i((k+1)T) \\ \bar{y}_i((k+1)T) \end{bmatrix} = \begin{bmatrix} \bar{x}_i(kT) + \int_{kT}^{(k+1)T} \bar{v}_i(t) \cos(\theta_i(t)) dt \\ \bar{y}_i(kT) + \int_{kT}^{(k+1)T} \bar{v}_i(t) \sin(\theta_i(t)) dt \end{bmatrix} + w_i(kT) \qquad (4.38)$$

with $\bar{x}_i(0) = x_i(0)$ and $\bar{y}_i(0) = y_i(0)$, and the vector $w_i(kT)$ defined by

$$w_i(kT) := 2v_0T \times \int_{(k+1/2)T}^{(k+1)T} \begin{bmatrix} \cos(\theta_i(t)) - \cos(\theta_0) \\ \sin(\theta_i(t)) - \sin(\theta_0) \end{bmatrix} dt.$$

Theorem 4.2 *Consider n mobile sensors described by equation (4.3). Suppose that Assumptions 4.1–4.4 are satisfied. Then the decentralized control law (4.28), (4.29), and (4.37) is an optimal corridor-sweep-coverage coordinated control with sweeping speed v_0 for the mobile sensors.*

Proof: Based on (4.3) and (4.38), we can write $(x_i(kT), y_i(kT))$ as follows:

$$\begin{bmatrix} x_i(kT) \\ y_i(kT) \end{bmatrix} = \begin{bmatrix} \bar{x}_i(kT) + \hat{x}(kT) \\ \bar{y}_i(kT) + \hat{y}(kT) \end{bmatrix} \tag{4.39}$$

for $i = 1, 2, \ldots, n$, where $(\hat{x}(kT), \hat{y}(kT))$ is governed by

$$\begin{aligned} \hat{x}((k+1)T) &= \hat{x}(kT) + v_0T\cos(\theta_0), \\ \hat{y}((k+1)T) &= \hat{y}(kT) + v_0T\sin(\theta_0) \end{aligned} \tag{4.40}$$

with $\hat{x}(0) = \hat{y}(0) = 0$.

Assumption 2.4 and the update rule (4.29) guarantee that $\lim_{k\to\infty}\theta_i(kT) = \theta_0$. In turn, the vector $w_i(kT)$ in (4.38) satisfies $\lim_{k\to\infty}\|w_i(kT)\| = 0$ for $i = 1, 2, \ldots$. By using Theorem 4.1 in Section 4.4 and the input-to-state-stability property of the consensus algorithm [92], it can be shown that the decentralized control law (4.28) and (4.29) is an optimal corridor-barrier-coverage coordinated control for the system (4.38). Then, there exists a line (4.31) such that $\lim_{k\to\infty} d(\bar{p}_i(kT), \mathscr{L}_0) = 0$, where $\bar{p}_i(kT) = (\bar{x}_i(kT), \bar{y}_i(kT))^T$ and $i = 1, 2, \ldots, n$. Also, there exists a permutation $\{z_1, z_2, \ldots, z_n\}$ of the set $\{1, 2, \ldots, n\}$ such that

$$\lim_{k\to\infty} \|\bar{p}_{z_i}(kT) - h_i\| = 0, \quad i = 1, 2, \ldots, n, \tag{4.41}$$

where h_i is defined by (4.9).

From (4.12) and (4.13), we get $h_i(kT) = h_i + kTv_0 \times [\cos(\theta_0)\ \sin(\theta_0)]^T$, $i = 1, 2, \ldots, n$. Along with equations (4.39)–(4.41), this gives

$$\lim_{k\to\infty} \|p_{z_i}(kT) - h_i(kT)\| = \lim_{k\to\infty} \|\bar{p}_{z_i}(kT) - h_i\| = 0 \tag{4.42}$$

for $i = 1, 2, \ldots, n$ since $[\hat{x}_i(kT)\ \hat{y}_i(kT)]^T = kTv_0 \times [\cos(\theta_0)\ \sin(\theta_0)]^T$. Hence, condition (4.14) is satisfied, which completes the proof of Theorem 4.2. ∎

4.6 Illustrative Examples

Now we present some simulation results to illustrate the proposed algorithm. First, we consider six mobile sensors and employ them for barrier coverage in a straight corridor. Figure 4.3(a) shows their positions at the initial deployment. Initially, they are stationary and the arrows indicate their headings. The locations of the sensors driven by the proposed algorithm are shown in Fig. 4.3(b) at time $t = 100T$, where $T = 1$. As can be seen, the sensors converge to a stationary line that is orthogonal to the corridor. Moreover, the sensors are evenly distributed over the line segment between the walls of the corridor. Since the sensing range is $r = 2$, the sensors form a barrier that can detect any passing intruder. Next, we employ the same six mobile sensors for sweep coverage in the same corridor, and their initial positions in this experiment are shown in Fig. 4.4(a). Our proposed sweep coverage algorithm steers the sensors into a line formation, and they maintain this formation while sweeping along the corridor. As a result, any stationary intruder or object with ordinate $y \geq 1$ is detected. Here the sensors sweep along the corridor at a speed of $v_0 = 0.09$.

Our proposed algorithm was developed under the assumption that the corridor is straight. However, the algorithm can also be applied to smooth curved corridors. For example, Figs. 4.5 and 4.6 show the simulation results of barrier coverage in a circular corridor and in an arbitrarily curved corridor. Sweep coverage can also be achieved in these corridors, as is shown in Figs. 4.7 and 4.8. As in the case of a straight corridor, the sensing ranges $r = 2$ and $R = 3.5$ were used. For the circular corridor, we chose $v_0 = 0.2$, whereas $v_0 = 0.6$ for the arbitrarily curved corridor. However, v_0 is not the actual sweeping speed along a curved corridor. This is because the forward speed of the sensors are different as they maneuver along a curved corridor.

So far we have showed that our control algorithm is robust against certain classes of smooth corridors. In the next simulation test, we explore the sensitivity of the algorithm to uncertainty in position measurements and sensor malfunction. Figure 4.9(a) shows a simulation result of barrier coverage with $r = 2$, $R = 3.5$, and bounded noises in the measurements of the positions of neighboring sensors: The positional readings were corrupted by independent additive noises uniformly distributed over $[-0.2, 0.2]$. As is shown in Fig. 4.9(a), a sensor barrier is indeed formed, and this sensor barrier can detect any intruder that moves across it since the distance between each pair of adjacent sensors is less than r. However, this sensor barrier does not converge to a straight line and the sensors in the barrier in fact oscillate within a bounded range. Nevertheless, it serves the purpose of barrier coverage. Figure 4.9(b) shows a simulation of sweep coverage with $v_0 = 0.09$ and the above bounded position measurement noises. Even though the sensors are not able to converge to a moving straight line, they are still able to move along the corridor in a formation close to a straight line, with the distance between adjacent sensors being less than r. Hence, this moving sensor barrier again provides sweep coverage along the corridor.

Interestingly, Figs. 4.10(a)–(b) show that if the accuracy of the position measurement improves as time progresses and the measurement noise eventually vanishes, both barrier and sweep coverages are achieved in a way similar to that observed in the respective cases with perfect position measurements (cf. Definitions 4.1 and 4.2). The simulations shown in Figs. 4.10(a)–(b) are generated using vanishing additive position measurement noises $\Delta x(kT)$ and $\Delta y(kT)$ that are described by

$\Delta x(kT) = \eta_x(kT)(0.9)^k$ and $\Delta y(kT) = \eta_y(kT)(0.9)^k$, respectively, where $\eta_x(kT)$ and $\eta_y(kT)$ are uniformly distributed over $[-1, 1]$.

Figures 4.11(a)–(b) show simulations of barrier and sweep coverages, respectively, with a sensor malfunction. In both cases, once a sensor failed, the functioning sensors reorganize and recover the barrier and sweep coverages. In other words, the algorithm is robust against sensor malfunction, provided that there are sufficiently many redundant sensors in the network.

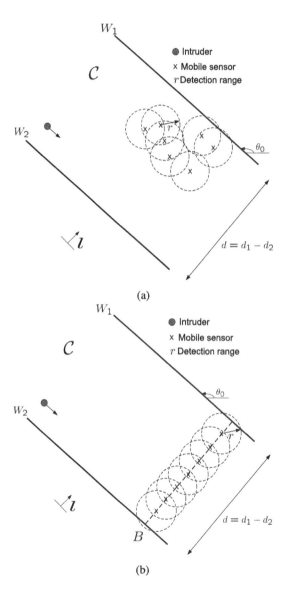

Figure 4.1   (a) Initial deployment with an intruder undetected; (b) final deployment with the intruder detected.

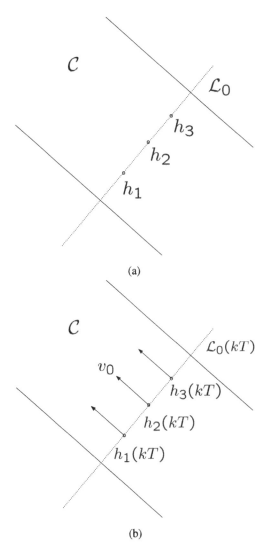

(a)

(b)

Figure 4.2 (a) Desired sensor locations for barrier coverage ($n = 3$); (b) Desired sensor trajectories for sweep coverage ($n = 3$).

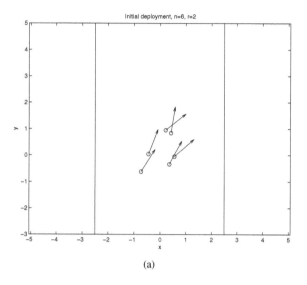

(a)

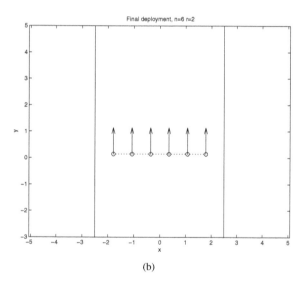

(b)

Figure 4.3   Barrier coverage in a $90°$ straight-line corridor ($r = 2$, $R = 3.5$): (a) Initial deployment; (b) Final deployment.

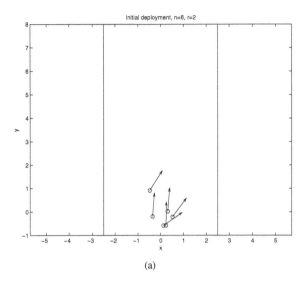

(a)

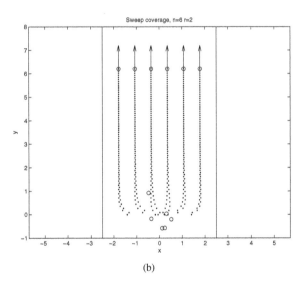

(b)

Figure 4.4   Sweep coverage within a 90° straight-line corridor ($r = 2$, $R = 3.5$, $v_0 = 0.09$): (a) Initial deployment; (b) The trajectories of the sensors.

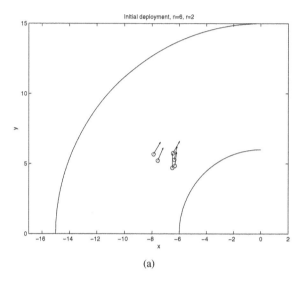

(a)

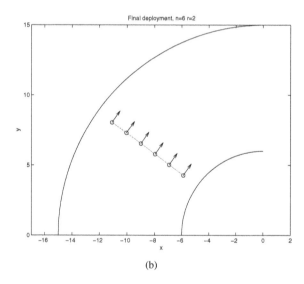

(b)

Figure 4.5   Barrier coverage in a circular corridor ($r = 2$, $R = 3.5$): (a) Initial deployment; (b) Final deployment.

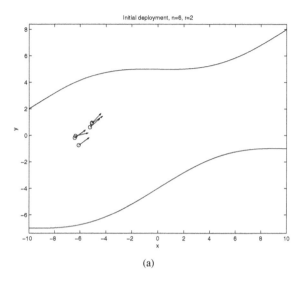

(a)

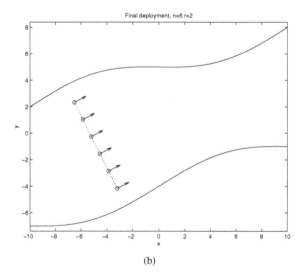

(b)

Figure 4.6 Barrier coverage in an arbitrarily curved corridor ($r = 2$, $R = 3.5$): (a) Initial deployment; (b) Final deployment.

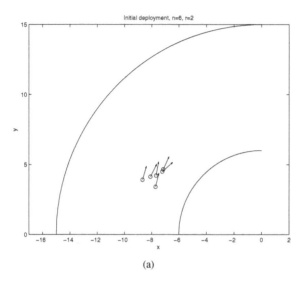

(a)

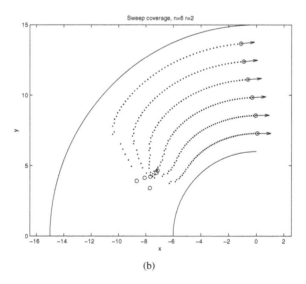

(b)

Figure 4.7   Sweep coverage with a circular corridor ($r = 2$, $R = 3.5$, $v_0 = 0.2$): (a) Initial deployment; (b) The trajectories of the sensors.

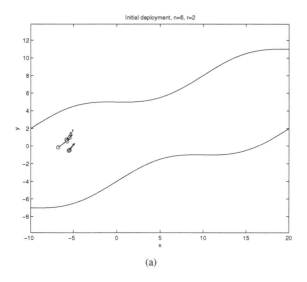

(a)

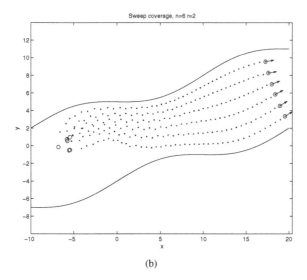

(b)

Figure 4.8 Sweep coverage with an arbitrarily curved corridor ($r = 2$, $R = 3.5$, $v_0 = 0.6$): (a) Initial deployment; (b) The trajectories of the sensors.

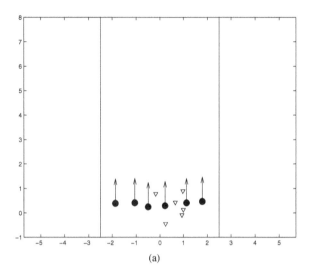

(a)

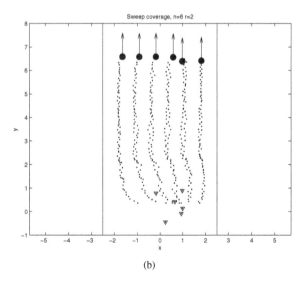

(b)

Figure 4.9 (a) Barrier coverage with bounded random position measurement noise (∇— initial position; ●—final position); (b) Sensor trajectories of sweep coverage with bounded random position measurement noise (∇—initial position).

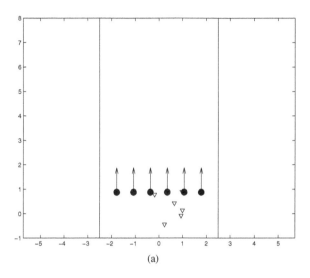

(a)

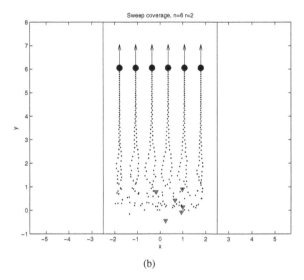

(b)

Figure 4.10   (a) Barrier coverage with vanishing random position measurement noise (∇—initial position; •—final position); (b) Sensor trajectories of sweep coverage with vanishing random position measurement noise (∇—initial position).

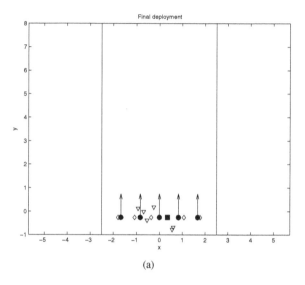

(a)

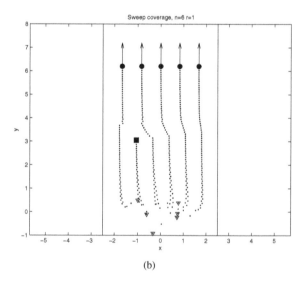

(b)

Figure 4.11 (a) Barrier coverage with a failed sensor (∇—initial position; ■—failed sensor; $\Diamond$—sensor position before a sensor failed; •—final position); (b) Sensor trajectories of sweep coverage with a failed sensor (∇—initial position; ■—failed sensor).

CHAPTER 5

SWEEP COVERAGE ALONG A LINE

5.1 Introduction

One of the key issues in research in sensor networks is that of coverage. In this context, coverage is typically treated as a measure of the network performance in surveillance of the physical space. In [35], Gage introduced an important type of coverage concerned with mobile sensor networks, which has further attracted an extensive research. This is the *sweep coverage*, which, according to Gage, means that the sensors are driven across the sensed field in a way that provides a balance between maximizing the detection rate of events and minimizing the number of missed detections per unit area.

Missions where sweep coverage is a real concern include, but are not limited to, maintenance inspection, reconnaissance, ship hull cleaning, and multi-sensor mine sweeping [35], as well as robot navigation for environmental extremum seeking [28, 78] and environmental field level tracking [79]. Among them, minesweeping is a drastically challenging and dangerous task [1, 12, 36], where the use of unmanned robotic devices is urgent. In view of this, the current chapter is focused on sweep coverage by deploying a network of autonomous robots that resembles a minesweeping operation. For cost reduction reason, the robots in the network are often endowed

Decentralized Coverage Control Problems for Mobile Robotic Sensor and Actuator Networks. **57**
By Andrey V. Savkin, Teddy M. Cheng, Zhiyu Xi, Faizan Javed, Alexey S. Matveev, and Hung
Nguyen. Copyright © 2015 by the Institute of Electrical and Electronics Engineers, Inc.

only with limited communication capacities, which means that they should work, either entirely or mostly, in an unsupervised and autonomous mode. As a result, motion control of such networks falls into the general framework of decentralized control, so far as every robot has access to only local information.

To address the aforementioned issues, this chapter develops several simple decentralized control laws that autonomously drive a team of robotic mobile sensors to form a sensor barrier, with maintaining a prespecified speed of motion. The objective is to steer the robots along a given line so that every point in some neighborhood of the line is detected by some sensor node. Therefore, the concerned problem can be referred to as *sweep coverage along a given line*. This problem is somewhat different from the sweep coverage problem in corridor environments, which was the subject of the previous chapter.

Besides minesweeping, plausible applications of the sweep coverage along a line include, e.g., border patrolling [59], sea floor surveying for hydrocarbon exploration [9], and environmental monitoring of disposal sites on the deep ocean floor [54]. In the case of sea exploration, the interest to the sweep coverage problem is, to some extent, inspired by the United Nations Convention on the Law of the Sea [118]. This convention defines the Exclusive Economic Zone (EEZ), where a coastal nation has sole exploitation rights over all natural resources, as an area that starts at the coastal baseline and extends 200 nautical miles (370.4 km) out into the sea. To accomplish exploration of a nation's EEZ, it suffices to deploy many low-cost, sensor-equipped underwater and/or surface vessels, and to let them sweep the EEZ as they move along the coastline. With regard to the magnitude of operation, the use of autonomous robotic vessels is a reasonable option, which basically puts the scenario in the framework of sweep coverage studied in this chapter.

The sweep coverage problems previously studied in, e.g., [26, 38, 60] are associated with coverage path-planning based on a more or less comprehensive map of the environment. This map is built from a priori and sensory data and used by the path planner for guiding the robots to cooperatively cover a target region. The underlying path planning algorithms are often endowed with both autonomous and cooperative functionalities [8, 82], with the main task for the latter being avoidance of collisions between the robots and improvement the overall efficiency of the coverage process. Two popular measures of this efficiency are as follows: (1) the time required for complete coverage and (2) the amount of repeated coverage. According to Zheng *et al.* [136], optimal multi-robot path planning for minimizing the coverage time is an NP-complete problem, so they offered polynomial-time coverage heuristics. Ge and Fua [41] examined the bounds of both the coverage time and amount of repeated coverage. The use of multiple robots for coverage promotes not only efficiency but also robustness, as is shown by, e.g., Hazon *et al.* [47]: They proposed a coverage algorithm that guarantees robustness against robot failures. Cooperation among the robots needs proper communication; in the context of coverage path planning, various relevant communication schemes are discussed in, e.g., [91, 121].

Compared with coverage path-planning, the sweep coverage problem addressed in this chapter is closer to cooperative multi-agent formation and marching control, where multiple robots are required to form up and move in a specified formation;

see, e.g., [11] and references therein. Similarly in our sweep coverage problem, a team of autonomous robotic sensors should be deployed to form a sensor barrier and to move along a given line while maintaining a desired barrier pattern.

Decentralized coordination and control of groups of autonomous robots becomes an emerging area of active research; see, e.g., [39, 89, 92] and references therein. In this area, a popular approach is based on the use of consensus algorithms; see, e.g., [53, 92, 96, 115, 127, 131]. Their core idea is to drive the information states or coordination variables of the team members to a common value (more generally, a set of values), i.e., to drive them to a consensus on this state or variable. Many available consensus algorithms employ the so-called "nearest neighbor rules" inspired by the animal aggregations [120]. According to these rules, each robot is controlled on the basis of information about the coordinates and/or velocities of only its closest neighbors — more precisely, the companion robots that are closest to the robot at hand at the current time. In this chapter, we follow this approach and develop several decentralized control laws for a group of autonomous mobile sensors to address the sweep coverage along a line.

The review paper [39] provides a detailed insight into the available algorithms of multi-agent coordination from the system dynamics and control perspective and discusses most widely adopted approaches to formation control, such as potential-based and behavior-based methods. The former introduces a certain artificial potential function, which takes into account the control objective and the interactions of every agent with both its companions and environment, whereas the system is driven along the negative gradient of this function. As discussed in [39], the major drawback of this approach is that in face of local minima of the potential function, which cannot be excluded in general, the desired formation pattern may not be guaranteed. Somewhat similarly, it is typically difficult to provide mathematically rigorous guarantees of the convergence to the desired formation for the alternative, behavior-based approach. Unlike these approaches, the algorithm presented in this chapter is supplied with mathematical guarantees that the mobile sensors come into the desired formation pattern. Das *et al.* [30] offered an extended study of formation control of a group of non-holonomic robots, which was based on input–output linearization and the use of a switching strategy. However, this approach may be troubled if connections between the leader and followers are lost, so that formation stabilization cannot be guaranteed. Conversely, the approach adopted in this chapter admits such a loss of connections for some times, still providing formation stabilization. Another distinction is that this approach stems from the idea of information consensus. An interesting multi-sensor formation problem is studied in Hayakawa *et al.* [46]. They proposed an energy-based controller, assuming that every sensor communicates with other sensors at discrete-time instances, and the communication graph is connected and does not vary over time. In contrast, this chapter assumes a time-varying communication graph and admits that it may be disconnected for some periods of time.

This chapter is based on results originally published in [25].[1]

The remainder of the chapter is organized as follows. In Section 5.2, we formulate the problem of decentralized control of autonomous robots for sweep coverage along a line. We then develop a control algorithm to solve this problem in Section 5.3. Section 5.4 presents some simulation results to demonstrate the effectiveness of the proposed algorithm. Finally, Section 5.5 contains proofs of some technical facts used to justify the main result of this chapter.

5.2   Problem of Sweep Coverage along a Line

We consider a planar straight line W with the slope $\bar{\phi}$:

$$W := \left\{ p \in \mathbb{R}^2 : l^T p = d_1 \right\}, \quad \bar{\phi} := \beta + \pi/2. \tag{5.1}$$

Here $l = [\cos(\beta) \ \sin(\beta)]^T$ is the unit vector normal to the line, $\beta \in [-\pi/2 \ \pi/2)$ is the slope of this vector with respect to the abscissa axis, and d_1 is a given constant.

The plain hosts a group of autonomous sensor-equipped mobile robots or vehicles. They should detect targeted objects, called *landmines* for the definiteness, in a region that starts at the line W and extends from it to a specified distance $L > 0$. All vehicles should operate on one side of the line W (for instance, surface/underwater vehicles can move only on/in the land/water during EEZ exploration). To accomplish the mission, each vehicle is equipped with a sensor that can detect landmines within a range of $s/2 > 0$. The initial distribution of the robots is random and on the correct side of W, but does not necessarily ensure a complete detection coverage of the region; see Fig. 5.1(a). Moreover, it is tacitly assumed that this region extends far along the line so that the number of sensors is not enough for its complete static coverage.

For this scenario, a way to achieve the detection objective is that the vehicles form a straight-line sensor barrier with a length L from W, which is perpendicular to W, and this sensor barrier sweeps the specified region as they move along W. Ideally, the vehicles are to be evenly spaced by the distance s to maximize the length L for a given number of vehicles; see Fig. 5.1(b). These goals should be achieved in a decentralized fashion: Every vehicle should operate autonomously and only on the basis of its own sensory data and information received via communication with companion vehicles that are currently within a *communication range* of $\rho > 0$. The sensory capacity of the vehicle includes the ability to detect W within a range of $R > 0$ and to estimate the slope $\bar{\phi}$ of W.[2] The vehicles are anonymous to one another: No vehicle can distinguish between any two other teammates.

Now we come to details. We consider n autonomous vehicles, labelled 1 through n. The symbols $x_i(\cdot), y_i(\cdot), v_i(\cdot)$, and $\theta_i(\cdot)$ stand for the abscissa, ordinate, linear

---

[1]The figures of this chapter are reprinted from Cheng T. M., Savkin A. V., and Javed F.: Decentralized control of a group of mobile robots for deployment in sweep coverage. Robotics and Autonomous Systems. **59(7-8)**, 497-507 (2011). Copyright ©2011 Elsevier. Reprinted with permission from Elsevier.

[2]The slope $\bar{\phi}$ can be estimated by finding the shortest distance from the vehicle to W.

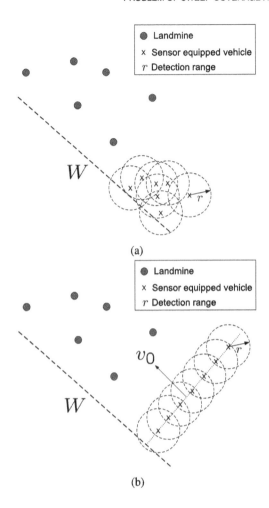

Figure 5.1   (a) Initial deployment; (b) Sweep coverage along the wall W.

velocity, and heading of vehicle i in the world frame, respectively. (The heading is measured from the abscissa axis in the counterclockwise direction.) The vehicle is considered as a self-propelled particle with a holonomic drive mechanism, so that the velocity v_i and heading θ_i are the control inputs of vehicle i. The first of them satisfies $|v_i(t)| \leq v_{max}$ for all $t \geq 0$, where $v_{max} > 0$ is common for all $i = 1, 2, \ldots, n$.

The discrete-time kinematic equations of vehicle i are given by

$$
\begin{aligned}
x_i((k+1)T) &= x_i(kT) + v_i(kT)T\cos(\theta_i(kT)), \\
y_i((k+1)T) &= y_i(kT) + v_i(kT)T\sin(\theta_i(kT)).
\end{aligned}
\tag{5.2}
$$

Here $k = 0, 1, 2, \ldots$ and $T > 0$ is a given sampling period: At discrete time instances $t = kT$, every vehicle gathers information about its neighborhood, communicates

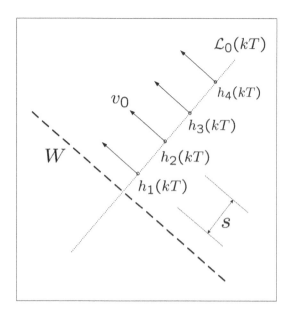

Figure 5.2   Desired vehicles' trajectories for sweep coverage ($n = 4$).

with other vehicles within a range of $\rho > 0$, and makes the decision about its own displacement during the time interval from kT to $(k+1)T$.

To complete the problem statement, we denote by $\mathscr{S}_1 := \{p \in \mathbb{R}^2 : l^T p > d_1\}$ and $\mathscr{S}_2 := \{p \in \mathbb{R}^2 : l^T p < d_1\}$ the two sides of the line (5.1) and denote by $\mathscr{S}_{\text{oper}}$ the side where the vehicles should operate. We also pick a desired sweeping speed

$$v_0 \in (0, v_{\max}) \tag{5.3}$$

and a scalar constant $\bar{c}$, and we define a line $\mathscr{L}_0(kT)$ that is perpendicular to the primary line (5.1) and moves at the speed v_0:

$$\mathscr{L}_0(kT) := \left\{ p = (x, y)^T \in \mathbb{R}^2 : x \cdot \cos \bar{\phi} + y \cdot \sin \bar{\phi} = \bar{c} + kT v_0 \right\} \tag{5.4}$$

for $k = 0, 1, 2, \ldots$. We also introduce $(n+1)$ points $h_i(kT)$ on this line (see Fig. 5.2), where $h_0(kT) := \mathscr{L}_0(kT) \cap W$ is the intersection point of $\mathscr{L}_0(kT)$ with the primal line W, and the other points lie on the "operational" side of W and are evenly spaced by a distance s, with the first of them being at the distance $s/2$ from W:

$$h_i(kT) = \begin{cases} h_0(kT) + \frac{s}{2}l + [s \cdot (i-1)]l & \text{if } \mathscr{S}_{\text{oper}} = \mathscr{S}_1, \\ h_0(kT) - \frac{s}{2}l - [s \cdot (i-1)]l & \text{if } \mathscr{S}_{\text{oper}} = \mathscr{S}_2 \end{cases} \tag{5.5}$$

for $i = 1, 2, \ldots, n$ and $k = 0, 1, 2, \ldots$. It is easy to see that the line $\mathscr{L}_0(kT)$, along with its points $h_i(kT)$, moves in the direction of $\bar{\phi}$ at the desired speed v_0. In fact, the points $h_i(kT), i = 1, \ldots, n$ constitute an "ideal" sensing barrier that runs at the required speed over the desired region and ensures its complete detection coverage.

Remark 5.1 *The subsequent figures depict not the line W but its shift by $s/2$ in the direction of $-l$ if $\mathscr{S}_{oper} = \mathscr{S}_1$, and in the opposite direction otherwise. This brings the convenience of equalizing all key distances concerned with the barrier (including the distance from $h_1(kT)$ to the line).*

Definition 5.1 (Sweep Coverage) *A set of decentralized control laws (each driving a particular robot) is said to ensure a sweep coverage at the speed v_0 along the line W and with the equidistance of s between the robots if for almost all initial positions of the robots, there exists a constant $\bar{c}$ and a permutation $\{z_1, z_2, \ldots, z_n\}$ of the set $\{1, 2, \ldots, n\}$ of robots such that the following convergences hold:*

$$\|p_{z_i}(kT) - h_i(kT)\| \to 0 \quad \text{as} \quad k \to \infty, \quad i = 1, 2, \ldots, n. \tag{5.6}$$

Here and throughout this chapter, $p_j(kT) = [x_j(kT), y_j(kT)]^T$ is the position of vehicle j at time $t = kT$.

Remark 5.2 *In (5.5), s was set to be twice the detection range of the robots in order to maximize the width of the covered area. At the expense of lessening this width, lesser s can be employed in (5.5); then some landmines from the covered area are multiply detected by the sensing barrier (5.5). From now on, $s > 0$ will be treated as a free parameter of the algorithm, which should not exceed twice the true detection range of the robots.*

The next section presents a control strategy that ensures a coverage described in Definition 5.1.

5.3 Sweep Coverage along a Line

For cost effective reason, the autonomous vehicles may have very limited communication and detection capacities and should be equipped with distributed or decentralized control laws that drive every vehicle on the basis of information about only its neighbors and itself. In this section, we offer a set of such control rules for coordination of the vehicles to achieve sweep coverage along the given line. These rules stem from deterministic distributed consensus algorithms.

The proposed control strategy assumes that every vehicle i memorizes and at times $t = kT$ updates a "coordination variable" $\phi_i(\cdot)$ with the meaning of "the opinion of this vehicle on the slope of the line W." The value of this variable is stamped with either "true" or "questionable" by coupling with 1 and 0, respectively. The initial value of $\phi_i(\cdot)$ is marked as "questionable" and is arbitrarily chosen; the choices of different vehicles are not correlated. At any time kT, vehicle i broadcasts this stamped value $\phi_i(kT)$, along with its stamp and the coordinates $x_i(kT)$ and $y_i(kT)$, and, in return, receives similar data from the companion vehicles lying within its communication range ρ. To avoid intricacies with the algorithm behavior, vehicle i accepts only the messages from a square *communication zone* $\mathscr{Z}_i(kT)$. This is the quadrate centered at this vehicle whose side has a prespecified length of $2r \leq \sqrt{2}\rho$ and is aligned with $\phi_i(kT)$. These data will be used to coordinate the motion of vehicle i with the neighbors, and ultimately with the entire team.

Definition 5.2 *The vehicles that are in the communication zone of vehicle i at time* $t = kT$ *are called its* neighbors *at this time.*

To state the rule to update the variable $\phi_i(\cdot)$ at time $t = kT$, we denote by $\mathscr{N}_i(kT)$ the set of all neighbors $j, j \neq i$ of vehicle i at this time and denote by $\mathscr{N}_i^{\text{true}}(kT) := \{j \in \mathscr{N}_i(kT) \cup \{i\} : \phi_j(kT)$ is marked as "true"$\}$ the set of the neighbors, including itself, whose messages are stamped as "true." (Any of these sets may be empty.) We also denote by $\|\cdot\|$ the Euclidean norm and introduce the disk within which vehicle i can detect the line W and estimate its slope $\bar{\phi}$ at time $t = kT$:

$$D_{i,R}(kT) := \{p \in \mathbb{R}^2 : \|p - p_i(kT)\| \leq R\}. \tag{5.7}$$

Vehicle i updates $\phi_i(\cdot)$ in accordance with the following rules:

(1) If $D_{i,R}(kT) \cap W \neq \emptyset$, the coordination variable takes the value of the true slope $\phi_i((k+1)T) := \bar{\phi}$ of the baseline W and is marked as "true";

(2) Otherwise $D_{i,R}(kT) \cap W = \emptyset$, the algorithm operates as follows:

    (a) If $\mathscr{N}_i^{\text{true}}(kT) \neq \emptyset$, an index $j \in \mathscr{N}_i^{\text{true}}(kT)$ is selected, $\phi_i((k+1)T) := \phi_j(kT)$, and $\phi_i((k+1)T)$ is stamped with "true";

    (b) If $\mathscr{N}_i^{\text{true}}(kT) = \emptyset$, vehicle i averages the current values of the coordination variable over the neighbors and itself:

$$\phi((k+1)T) := \frac{1}{1 + |\mathscr{N}_i(kT)|} \sum_{j \in \mathscr{N}_i(kT) \cup \{i\}} \phi_j(kT). \tag{5.8}$$

Here $|S|$ stands for the number of elements in a finite set S (0 if $S = \emptyset$). The updated value $\phi((k+1)T)$ is marked as "questionable".

Remark 5.3 *It is easy to see that the choice of* $j \in \mathscr{N}_i^{\text{true}}(kT)$ *in 2.a) does not affect the overall result since* $\phi_i[(k+1)T]$ *is set to the true slope* $\bar{\phi}$ *under any choice.*

To underscore that $\phi_i((k+1)T)$ is computed at time kT and so can be used in other computations carried out at this time, we will also employ the alternative notation

$$\mathscr{H}_i(kT) = \phi_i((k+1)T). \tag{5.9}$$

To generate the current controls $v_i(kT)$ and $\theta_i(kT)$, vehicle i first uses $\mathscr{H}_i(kT)$ to calculate the following scalars:

$$c_{ij}(kT) = \begin{bmatrix} \cos(\mathscr{H}_i(kT)) & \sin(\mathscr{H}_i(kT)) \end{bmatrix} \begin{bmatrix} x_j(kT) \\ y_j(kT) \end{bmatrix}, \tag{5.10}$$

where j ranges over $\mathcal{N}_i(kT) \cup \{i\}$. This calculation is feasible since vehicle i has access to all concerned coordinates $x_j(\cdot), y_j(\cdot)$. The quantity (5.10) can be interpreted as the projection of the position of vehicle j onto the line passing through the origin with the slope $\mathcal{H}_i(kT)$. Similarly to (5.8), the "projections" $c_{ij}(\cdot)$ are averaged:

$$\mathcal{M}_i(kT) := \frac{1}{1 + |\mathcal{N}_i(kT)|} \sum_{j \in \mathcal{N}_i(kT) \cup \{i\}} c_{ij}(kT). \tag{5.11}$$

After this, we define the line $\mathcal{L}_i(kT)$ that is perpendicular to the updated slope (5.9) and passes through the averaged projection:

$$\mathcal{L}_i(kT) = \{(x,y) \in \mathbb{R}^2 : x \cos(\mathcal{H}_i(kT)) + y \sin(\mathcal{H}_i(kT)) = \mathcal{M}_i(kT)\}. \tag{5.12}$$

The vehicle will be moved to this line preliminarily translated by $v_0 T$ in its normal direction. To specify its future position on this line, vehicle i first determines the "projection" of the position of every vehicle $j \in \mathcal{N}_i(kT) \cup \{i\}$ on the line $\mathcal{L}_i(kT)$:

$$q_j^i(kT) = \begin{bmatrix} \sin(\mathcal{H}_i(kT)) & -\cos(\mathcal{H}_i(kT)) \end{bmatrix} \begin{bmatrix} x_j(kT) \\ y_j(kT) \end{bmatrix}. \tag{5.13}$$

Similarly if $D_{i,R}(kT) \cap W \neq \emptyset$, i.e., the vehicle is able to detect the baseline W, the "projection" of the point of intersection $\eta_i(kT) = \mathcal{L}_i(kT) \cap W$ is also computed:

$$q_0^i(kT) := \begin{bmatrix} \sin(\mathcal{H}_i(kT)) & -\cos(\mathcal{H}_i(kT)) \end{bmatrix} \eta_i(kT). \tag{5.14}$$

Among all so obtained "projections" $q_j^i(kT)$, the vehicle chooses the two that are closest to its own "projection" from the left and right, respectively,

$$q_\alpha^i(kT) < q_i^i(kT) < q_\beta^i(kT) \tag{5.15}$$

if such a couple does exist. If there is no "projection" to the left/right of $q_i^i(kT)$, the index α/β is undefined; both of them are undefined if there are no "projections" except for $q_i^i(kT)$. (Typically the last case holds if vehicle i has no neighbors and does not detect the line W.) The desired "projection" $\mathcal{Q}_i(kT)$ of the next position of vehicle i onto the line $\mathcal{L}_i(kT)$ is given by

$$\mathcal{Q}_i(kT) =$$

$$\begin{cases} \frac{1}{3}\left[q_\alpha^i(kT) + q_i^i(kT) + q_\beta^i(kT)\right] & \text{if both } \alpha \text{ and } \beta \text{ exist and } \alpha, \beta \neq 0; \\ \frac{1}{3}q_\alpha^i(kT) + \frac{1}{2}q_i^i(kT) + \frac{1}{6}q_\beta^i(kT) & \text{if both } \alpha \text{ and } \beta \text{ exist and } \alpha = 0; \\ \frac{1}{6}q_\alpha^i(kT) + \frac{1}{2}q_i^i(kT) + \frac{1}{3}q_\beta^i(kT) & \text{if both } \alpha \text{ and } \beta \text{ exist and } \beta = 0; \\ \frac{1}{2}\left[q_\alpha^i(kT) + q_i^i(kT) + s\right] & \text{if only } \alpha \text{ exists and } \alpha \neq 0; \\ \frac{1}{2}\left[q_\alpha^i(kT) + q_i^i(kT) + \frac{s}{2}\right] & \text{if only } \alpha \text{ exists and } \alpha = 0; \\ \frac{1}{2}\left[q_\beta^i(kT) + q_i^i(kT) - s\right] & \text{if only } \beta \text{ exists and } \beta \neq 0; \\ \frac{1}{2}\left[q_\beta^i(kT) + q_i^i(kT) - \frac{s}{2}\right] & \text{if only } \beta \text{ exists and } \beta = 0; \\ q_i^i(kT) & \text{if both } \alpha \text{ and } \beta \text{ do not exist.} \end{cases} \tag{5.16}$$

The specified projection $\mathcal{Q}_i(kT)$ onto $\mathcal{L}_i(kT)$, along with the intention to arrive at the line that results from $\mathcal{L}_i(kT)$ by the specified translation, in fact uniquely determines the next position $p_i[(k+1)T]$ of vehicle i.

It remains to find the controls $v_i(kT)$ and $\theta_i(kT)$ that drive the vehicle to this position for time T. To this end, we first determine the required (constant) speed in the direction of $\mathcal{L}_i(kT)$ and in the perpendicular direction, respectively:

$$\bar{v}_i(kT) := \frac{\mathcal{Q}_i(kT) - q_i^i(kT)}{T},$$

$$\hat{v}_i(kT) := \frac{\mathcal{M}_i(kT) - c_{ii}(kT) + v_0 T}{T}. \tag{5.17}$$

From this, the controls can be obtained in the following form:

$$v_i(kT) = \sqrt{\bar{v}_i(kT)^2 + \hat{v}_i(kT)^2},$$

$$\theta_i(kT) = \begin{cases} \phi_i((k+1)T) + \xi_i(kT) - \pi/2 & \text{if } \hat{v}_i(kT) \geq 0, \\ \phi_i((k+1)T) - \xi_i(kT) - \pi/2 & \text{if } \hat{v}_i(kT) < 0, \end{cases} \tag{5.18}$$

where $\xi_i(kT) := \cos^{-1}(\bar{v}_i(kT)/v_i(kT))$.

Thus the algorithm to control every vehicle is fully described. This algorithm is executable by anonymous robots: Since its result is invariant with respect to re-enumeration of the robots, as can be easily verified, their indices are not in fact required. Recognition of the index 0, which is needed to execute (5.16), is possible by our assumptions since this index is associated with not a robot but the event of detecting the baseline. Under minimal technical assumptions, feasibility of the generated control will be shown by Proposition 5.1 in the next section.

To illustrate the proposed control law, we consider vehicle i with three neighbors j_1, j_2, j_3, as is shown in Fig. 5.3. By using the coordination variables $\phi_i, \phi_{j_1}, \phi_{j_2}, \phi_{j_3}$ of vehicles i, j_1, j_2, j_3, vehicle i first obtains $\mathcal{H}_i(kT)$. Then it computes the projections $c_{i,i}, c_{i,j_1}, c_{i,j_2}$, and c_{i,j_3} of the positions p_i, p_{j_1}, p_{j_2}, and p_{j_3}, respectively, in the direction of the vector $\mathbf{a} = [\cos(\mathcal{H}_i(kT)) \quad \sin(\mathcal{H}_i(kT))]^T$; see Fig. 5.3. The quantity $\mathcal{M}_i(kT)$ is the average of $c_{i,i}, c_{i,j_1}, c_{i,j_2}, c_{i,j_3}$. Using $\mathcal{M}_i(kT)$ and $\mathcal{H}_i(kT)$, the line $\mathcal{L}_i(kT)$ is defined by (5.12). In (5.15), α and β correspond to the neighbors j_1 and j_2, as is shown in Fig. 5.3. By (5.13) and letting $\mathbf{b} = [\sin(\mathcal{H}_i(kT)) \quad -\cos(\mathcal{H}_i(kT))]^T$, we see that $q_\alpha^i = \mathbf{b}^T p_{j_1}$ and $q_\beta^i = \mathbf{b}^T p_{j_2}$. By (5.17) and (5.18), vehicle i determines the control $\mathbf{v}_i(kT)$, where $\mathbf{a}^T \mathbf{v}_i(kT) = \hat{v}_i(kT)$ and $\mathbf{b}^T \mathbf{v}_i(kT) = \bar{v}_i(kT)$. This control drives vehicle i to the next position $p_i((k+1)T)$ such that $p_i^T((k+1)T)\mathbf{a} = \mathcal{M}_i(kT) + v_0 T$ and $p_i^T((k+1)T)\mathbf{b} = \mathcal{Q}_i(kT)$.

Remark 5.4 *So far, it was tacitly assumed that each vehicle knows both its own location and the locations of the neighbors in the world frame. However, relative position measurements are enough.*

Indeed, let $\bar{p}_{i,j}(kT) = p_j(kT) - p_i(kT)$ stand for the relative position of vehicle j with respect to i, and similarly to (5.10),

$$\bar{c}_{ij}(kT) := \begin{bmatrix} \cos(\mathcal{H}_i(kT)) & \sin(\mathcal{H}_i(kT)) \end{bmatrix} \bar{p}_{i,j}(kT).$$

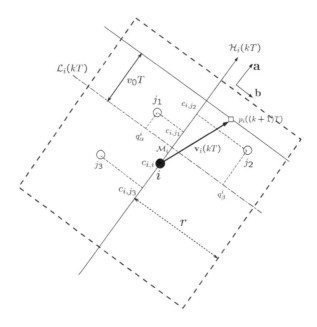

Figure 5.3   Explanation of the proposed algorithm.

Since $c_{ij}(kT) = \bar{c}_{ij}(kT) + c_{ii}(kT)$ by (5.10), relation (5.11) yields that

$$\mathscr{M}_i(kT) = \bar{\mathscr{M}}_i(kT) + c_{ii}(kT),$$

where

$$\bar{\mathscr{M}}_i(kT) := \frac{1}{1 + |\mathscr{N}_i(kT)|} \sum_{j \in \mathscr{N}_i(kT)} \bar{c}_{i,j}(kT).$$

It follows that the speed given by the second formula in (5.17) can be found based on the relative measurements $\bar{p}_{i,j}(kT)$ only:

$$\hat{v}_i(kT) = \frac{\bar{\mathscr{M}}_i(kT) + v_0 T}{T}.$$

To follow the same line of reasoning for the other speed $\bar{v}_i(kT)$ from (5.17), we copycat (5.13) by putting

$$\bar{q}^i_j(kT) := \Big[\sin(\mathscr{H}^o_i(kT)) \quad -\cos(\mathscr{H}^o_i(kT))\Big]\, \bar{p}_{ij}(kT)$$

and observe that $q^i_j(kT) = \bar{q}^i_j(kT) + q^i_i(kT)$ by (5.13). If $D_{i,R}(kT) \cap W \neq \emptyset$, i.e., the vehicle detects the line W, we also introduce the relative position $\bar{\eta}_i(kT) := \eta_i(kT) - p_i(kT)$ of the point of intersection $\eta_i(kT) = \mathscr{L}_i(kT) \cap W$ and note that $q^i_0(kT) = \bar{q}^i_0(kT) + q^i_i(kT)$ by (5.14). It follows that the numerator $\bar{\mathscr{Q}}_i(kT) := \mathscr{Q}_i(kT) - q^i_i(kT)$

in the first formula from (5.17) is given by the following analog of (5.16)

$$
\mathcal{Q}_i(kT) =
\begin{cases}
\frac{1}{3}\left[\bar{q}^i_\alpha(kT) + \bar{q}^i_i(kT) + \bar{q}^i_\beta(kT)\right] & \text{if both } \alpha \text{ and } \beta \text{ exist and } \alpha, \beta \neq 0; \\
\frac{1}{3}\bar{q}^i_\alpha(kT) + \frac{1}{2}\bar{q}^i_i(kT) + \frac{1}{6}\bar{q}^i_\beta(kT) & \text{if both } \alpha \text{ and } \beta \text{ exist and } \alpha = 0; \\
\frac{1}{6}\bar{q}^i_\alpha(kT) + \frac{1}{2}\bar{q}^i_i(kT) + \frac{1}{3}\bar{q}^i_\beta(kT) & \text{if both } \alpha \text{ and } \beta \text{ exist and } \beta = 0; \\
\frac{1}{2}\left[\bar{q}^i_\alpha(kT) + \bar{q}^i_i(kT) + s\right] & \text{if only } \alpha \text{ exists and } \alpha \neq 0; \\
\frac{1}{2}\left[\bar{q}^i_\alpha(kT) + \bar{q}^i_i(kT) + \frac{s}{2}\right] & \text{if only } \alpha \text{ exists and } \alpha = 0; \\
\frac{1}{2}\left[\bar{q}^i_\beta(kT) + \bar{q}^i_i(kT) - s\right] & \text{if only } \beta \text{ exists and } \beta \neq 0; \\
\frac{1}{2}\left[\bar{q}^i_\beta(kT) + \bar{q}^i_i(kT) - \frac{s}{2}\right] & \text{if only } \beta \text{ exists and } \beta = 0; \\
\bar{q}^i_i(kT) & \text{if both } \alpha \text{ and } \beta \text{ do not exist.}
\end{cases}
$$

Thus both $\hat{v}_i(kT)$ and $\bar{v}_i(kT)$, and so the entire current control (5.18), can really be written in terms of the relative positions.

Remark 5.4 gives rise to the following.

Remark 5.5 *The proposed control law can be implemented on the basis of various sensor and communication schemes.*

For example, it suffices that firstly, every robot has access to the absolute direction via, e.g., a compass and can determine orientation of its own local reference frame with respect to the world frame; and, secondly, sensors inform the robot about the relative position and orientation of the baseline (if it is within a range of R from the robot) and the relative positions of the teammates that are within a range of ρ from the robot at hand. In this case, only the stamped variables $\phi_j(kT)$ should be communicated: every vehicle i broadcasts its own variable to the neighbors and in return, receives their variables.

5.4 Assumptions and the Main Results

To perform analysis of the proposed algorithm, we need some assumptions.

Assumption 5.1 *The following inequalities hold:*

$$
s < \min\{r; 2R\}, \quad r \leq \frac{\rho}{\sqrt{2}}, \quad \sqrt{\max\left\{\frac{\rho}{3}, \frac{\rho-s}{2}, \frac{s}{2}\right\}^2 + \rho^2} < R,
$$

$$
\sqrt{\delta^2 + \rho^2} < v_{\max}T, \quad \text{where} \quad \delta := \max\left\{\frac{\rho}{3}, \frac{R}{3}, \frac{\rho-s}{2}, \frac{R-s/2}{2}\right\}.
\tag{5.19}
$$

Here the last inequality can always be satisfied by a proper choice of the sampling period T. The third inequality can be met via artificially reducing ρ if necessary (i.e., via ignoring messages from the vehicles at distances greater then the reduced ρ); for example, it is enough to ensure that $\rho < 2R/\sqrt{5}$ since $\max\left\{\frac{\rho}{3},\frac{\rho-s}{2},\frac{s}{2}\right\} \leq \frac{\rho}{2}$ under the first two conditions from (5.19). Invoking Remark 5.2, the first two inequalities can be satisfied by picking the parameters r and s of the algorithm small enough.

The desired speed of sweeping (5.3) is supposed to be known to every vehicle. The next assumption in fact offers not to desire an excessively large speed.

Assumption 5.2 *The sweeping speed $v_0 \neq 0$ is such that*

$$|v_0| \leq \frac{1}{T}\min\left\{v_{\max}T - \sqrt{\delta^2+\rho^2}; R - \sqrt{\max\left\{\frac{\rho}{3},\frac{\rho-s}{2},\frac{s}{2}\right\}^2 + \rho^2}\right\}. \quad (5.20)$$

Here the right-hand side is positive due to (5.19).

Now we show that under these assumptions, the proposed control law is feasible.

Proposition 5.1 *Let Assumptions 5.1, 5.2 hold. Then the following claims are true:*

(i) *The proposed control law generates a feasible control $|v_i(kT)| \leq v_{\max}$ at any time instant $t = kT$ for any vehicle i;*

(ii) *Under the proposed control strategy, the vehicles do not hit the line W and remain on the correct side of W.*

Proof: (i) Let us consider an arbitrary vehicle i and time instant $t = kT$. Due to Definition 5.2 and the second inequality from (5.19), any neighbor $j \in \mathcal{N}_i(kT)$ is separated from vehicle i by a distance no greater than the communication radius ρ. So by (5.10), $|c_{ij}(kT) - c_{ii}(kT)| \leq \rho$ either. Hence the mean (5.11) of the projections $c_{ij}(kT)$ also lies in the ρ-neighborhood $[c_{ii}(kT) - \rho, c_{ii}(kT) + \rho]$ of $c_{ii}(kT)$, and the second relation in (5.17) implies that

$$|\hat{v}_i(kT)| \leq (\rho + |v_0|T)/T. \quad (5.21)$$

By the same argument, the distance between $q_i^i(kT)$ and $q_j^i(kT)$ does not exceed ρ for $j \in \mathcal{N}_i(kT) \subset \{1,\ldots,n\}$. As for $j = 0$, the presence of the extra projection $q_0^i(kT)$ means that $D_{i,R}(kT) \cap W \neq \emptyset$ by construction, and so $\mathcal{H}_i(kT) = \bar{\phi}$ due to (5.9) and 1) on page 64. Hence by (5.13) (with $j = i$) and (5.14)

$$q_i^i(kT) - q_0^i(kT) = \left[\sin(\bar{\phi}) \quad -\cos(\bar{\phi})\right][p_i(kT) - \eta_i(kT)],$$

where $\eta_i(kT) \in W$ and the first multiplier on the right gives the unit vector perpendicular to W. It follows that $|q_i^i(kT) - q_0^i(kT)|$ does not exceed the distance from $p_i(kT)$ to W, which in turn, does not exceed R. Overall, $|q_i^i(kT) - q_0^i(kT)| \leq R$.

By (5.16), the foregoing implies that $|q_i^i(kT) - \mathcal{Q}_i(kT)|$ does not exceed

$$\max\left\{\frac{\rho}{3},\frac{R}{3},\frac{s}{2},\frac{\rho-s}{2},\frac{R-s/2}{2}\right\} = \max\left\{\frac{\rho}{3},\frac{R}{3},\frac{\rho-s}{2},\frac{R-s/2}{2}\right\} \overset{(5.19)}{=\!=\!=} \delta,$$

where the first equation holds since $(R - s/2)/2 \geq s/2$ due to the first inequality from (5.19). Hence (5.17) yields that

$$|\bar{v}_i(kT)| \leq \delta/T. \tag{5.22}$$

The proof of (i) is completed as follows:

$$v_i(kT) \stackrel{(5.18)}{=\!=\!=} \sqrt{\bar{v}_i(kT)^2 + \hat{v}_i(kT)^2} \leq \frac{1}{T}\sqrt{\delta^2 + (\rho + |v_0|T)^2}$$

$$\leq \frac{\sqrt{\delta^2 + \rho^2} + |v_0|T}{T} \stackrel{(5.20)}{\leq} v_{max}.$$

(ii) We argue by induction on k. For $k = 0$, the claim is a part of the assumptions about the initial locations of the vehicles. Let all vehicles be on the correct side of W at time $t = kT$. Suppose that the same is not true at time $t = (k+1)T$, i.e., the control law dictates some vehicle, say i, to cross W. We are going to show first that

$$D_{i,R}(kT) \cap W = \emptyset. \tag{5.23}$$

Indeed, suppose to the contrary that (5.23) does not hold. Then the extra projection $q_0^i(kT)$ is used, $\mathcal{H}_i(kT) = \bar{\phi}$ due to (5.9) and 1) on page 64, and so all projections (5.13) (with $j \in \mathcal{N}_i(kT) \cup \{i\}$) lie on a common side of the extra projection (5.14), which side corresponds to the correct side of W. It follows from (5.16) that $\mathcal{Q}_i(kT)$ is on the same side of $q_0^i(kT)$. So (5.17) and (5.18) entail transition of vehicle i to a position on the correct side of W, in violation of the initial hypothesis. The contradiction obtained proves (5.23).

Due to (5.23), the extra projection $q_0^i(kT)$ is not introduced for vehicle i at time $t = kT$. This and retracing the respective arguments from the proof of the part i) shows that now (5.22) takes the form $|\bar{v}_i(kT)| \leq T^{-1} \max\{\rho/3, (\rho - s)/2, s/2\}$ by dropping R, whereas (5.21) still holds. Hence for the updated location $p_i[(k+1)T]$, we have

$$\|p_i[(k+1)T] - p_i[kT]\| \leq v_i(kT)T \stackrel{(5.18)}{\leq} \sqrt{\max\left\{\frac{\rho}{3}, \frac{\rho-s}{2}, \frac{s}{2}\right\}^2 + (\rho + |v_0|T)^2}$$

$$\leq \sqrt{\max\left\{\frac{\rho}{3}, \frac{\rho-s}{2}, \frac{s}{2}\right\}^2 + \rho^2 + |v_0|T} \stackrel{(5.20)}{\leq} R.$$

However, this contradicts (5.23) since on the way from $p_i[kT]$ to $p_i[(k+1)T]$ vehicle i crosses W by the initial hypothesis. Thus this hypothesis is incorrect, and all vehicles remain on the correct side of W at time $t = (k+1)T$. They do not hit W during the time interval from kT to $(k+1)T$ since each of them moves along a straight line between two positions on the same side of W. ∎

To state the last assumption, we introduce two undirected graphs at any time instant $t = kT$. The first of them, $G(kT)$, is the standard communication graph visualizing the relation of neighborship: Its vertices are associated with the vehicles

$i = 1, \ldots, n$, and vertices $i \neq j$ are connected by an edge if and only if vehicles i and j are neighbors at time kT. The second graph $\bar{G}(kT)$ extends $G(kT)$. It contains one more vertex 0, retains all vertices and edges of the original graph $G(kT)$, and links the "new" vertex 0 with an "old" one $i \neq 0$ with an edge if and only if vehicle i detects the basic line W at time $t = kT$, i.e., (5.23) holds.

Assumption 5.3 *The both sequences of time-varying graphs $G(kT)$ and $\bar{G}(kT)$ (where $k = 0, 1, \ldots$) satisfy the standard Main Connectivity Assumption, i.e., Assumption 2.4 from Chapter 2 stated on page 11.*

Now we are in position to present the key result of this chapter.

Theorem 5.1 *Suppose that Assumptions 5.1–5.3 are satisfied. Then the set of the decentralized control laws (5.9) and (5.18) (where $i = 1, \ldots, n$) ensures a sweep coverage at the speed v_0 along the line W and with the equidistance of s between the robots in the sense of Definition 5.1.*

Proof: For the sake of convenience, the origin of the world frame is chosen on the baseline W so that $d_1 = 0$ in equation (5.1), which thus simplifies into

$$W = \{p \in \mathbb{R}^2 : l^T p = 0\}, \quad l = \begin{pmatrix} \sin \bar{\phi} \\ -\cos \bar{\phi} \end{pmatrix}. \tag{5.24}$$

Since the sequence $\bar{G}(\cdot)$ of graphs satisfies Assumption 5.3, algorithm 1), 2) from page 64 eventually drives the coordination variables of all vehicles at the value of the true slope $\bar{\phi}$ of the baseline W and after this, does not change this value any more: there exists $k_* = 0, 1, \ldots$ such that

$$\phi_i(kT) = \bar{\phi} \qquad \text{for all} \quad k \geq k_* \quad \text{and} \quad i = 1, \ldots, n. \tag{5.25}$$

From now on, we focus our analysis on $k \geq k_*$. Then due to (5.9) and (5.10),

$$\mathscr{H}_i^\circ(kT) = \bar{\phi}, \tag{5.26}$$

$$c_{ij}(kT) = c_j(kT) := x_j(kT) \cos \bar{\phi} + y_j(kT) \sin \bar{\phi}, \qquad j \in \mathscr{N}_i(kT) \cup \{i\}, \tag{5.27}$$

where the last equation underscores that $c_{ij}(kT)$ in fact does not depend on i. Since the origin of the world frame lies on W, formula (5.27) in fact determines the orthogonal projection of the current location of vehicle j onto W. The following claim will be proven in Section 5.6 based on the remaining part of Assumption 5.3 that is concerned with the graph $G(kT)$.

Claim 1. There exist a constant $\bar{c}$ such that for any vehicle i, the following holds:

$$c_i(kT) - (\bar{c} + v_0 Tk) \to 0 \quad \text{as} \quad k \to \infty. \tag{5.28}$$

In the light of (5.27), this claim means that the distance $d[p_i(kT), \mathscr{L}_0(kT)]$ from any vehicle i to the moving straight line (5.4) goes to zero, as is required:

$$d[p_i(kT), \mathscr{L}_0(kT)] \to 0 \quad \text{as} \quad k \to \infty \quad \text{for all} \quad i = 1, \ldots, n. \tag{5.29}$$

So by increasing k_* if necessary, it can be ensured that

$$d[p_i(kT), \mathcal{L}_0(kT)] < r/2, \qquad i = 1, \ldots, n, \quad k \geq k_*. \tag{5.30}$$

Then vehicles $i \neq j$ are in the communication zone of each other if and only if their orthogonal projections onto $\mathcal{L}_0(kT)]$ are separated by a distance no greater then r.

From now on, we assume that the vector $l = [\sin \phi \quad -\cos \phi]^T$ points to the "operational" side of the baseline W, for the sake of definiteness. (The converse case is considered likewise and so its detailed analysis is omitted.) Then the first row from (5.5) is in fact put in use. Furthermore, due to (5.26) and since the origin of the frame of reference is set on the baseline, $q_0^i(kT) = 0$ in (5.14). By the same argument and ii) from Proposition 5.1, the quantity $q_j^i(kT)$ given by (5.13) is positive, does not depend on i, i.e., $q_j^i(kT) = q_j(kT)$, and can be viewed as the orthogonal projection of $p_j(kT)$ onto the line (5.4). From now on, we consider the case where the projections $q_j(k_*T), j = 1, \ldots, n$, are pairwise different. Then after a proper re-enumeration (vehicle z_i takes the new label i, where $\{z_1, z_2, \ldots, z_n\}$ is a proper permutation of the set $\{1, 2, \ldots, n\}$), they may be put in a strictly ascending order: for $k := k_*$,

$$0 < q_1(kT) < q_2(kT) < \ldots < q_n(kT). \tag{5.31}$$

The next claim will be proven in Section 5.6.

Claim 2 Inequalities (5.31) remain true for all $k \geq k_*$.

Now we consider the moving points $h_i(kT) \in \mathcal{L}_0(kT), i = 1, \ldots, n, k = 0, 1, \ldots$, defined by the first row from (5.5). Similarly to (5.13), we introduce their orthogonal "projections" onto the line $\mathcal{L}_0(kT)$:

$$\mathfrak{h}_i(kT) = \left[\sin \bar{\phi} \quad -\cos \bar{\phi}\right] h_i(kT) = \begin{cases} 0 & \text{for } i = 0, \\ \frac{s}{2} + s(i-1) & \text{for } i = 1, \ldots, n. \end{cases}$$

To complete the proof of the theorem, it suffices to justify (5.6) with $z_i := i$. Since the points $h_i(kT)$ lie on the line $\mathcal{L}_0(kT)$ and (5.29) holds, (5.6) is equivalent to the fact that the distance between the orthogonal projections of $p_i(kT)$ and $h_i(kT)$ onto this line converges to zero as $k \to \infty$ for all $i = 1, \ldots, n$. Exactly this fact is established in the last claim, which thus completes the proof.

Claim 3. For any $i = 1, \ldots, n$, the following convergence holds:

$$|q_i(kT) - \mathfrak{h}_i(kT)| \to 0 \quad \text{as} \quad k \to \infty. \tag{5.32}$$

The proof of this claim is given in Section 5.6.   ∎

5.5 Illustrative Examples

In this section, the proposed algorithm is illustrated via computer simulations.

5.5.1   Straight-Line Sweeping Paths

First, we consider $n = 10$ vehicles and deploy them for a sweep coverage along a line W with the slope $\bar{\phi} = \pi/3$. The other parameters used for simulations are as follows: $r = 1, s = 0.5, R = 1.5, T = 1$, and $v_0 = 0.01$. Initially, the vehicles were stationary and scattered over the left-hand side of W.

Figure 5.4 shows the respective paths of the vehicles driven by the proposed control law and reveals achievement of the sweep coverage objective. Indeed, after a transient the vehicles form a sensor barrier that is orthogonal to W, and they move at the desired speed of $v_0 = 0.01$ in the desired direction of $\bar{\phi} = \pi/3$. Furthermore, the inter-vehicle distances and and the distance from the first of them to W (preliminarily shifted by $s/2$) converge to the desired value s so that any targeted object that lies within the distance $ns + s/2$ from W can be detected.

As is shown in Fig. 5.5, similar results are obtained for a line with the slope $\bar{\phi} = 3\pi/4$ and a team of twenty ($n = 20$) vehicles. The parameters for this simulation were chosen as $r = 1, s = 0.1, R = 1.5, T = 1$, and $v_0 = 1 \times 10^{-3}$.

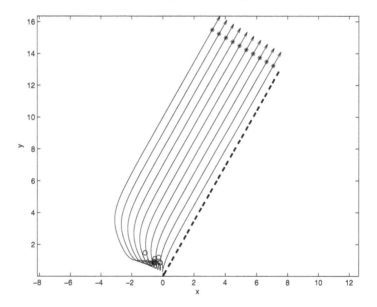

Figure 5.4   Sweep coverage with ten ($n = 10$) autonomous vehicles along the line W with the slope $\bar{\phi} = \pi/3$: initial positions (circles) and paths (solid lines) of the vehicles.

5.5.2   Comparison with the Potential Field Approach

To this end, a simulation was carried out where five vehicles were driven into a straight line formation based on an artificial potential field. The idea behind the employed field is to keep any two neighboring vehicles at a distance of 0.5 from

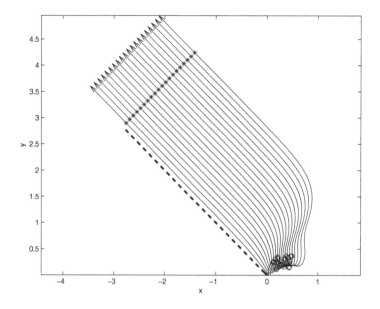

Figure 5.5   Sweep coverage with twenty ($n = 20$) autonomous vehicles along the line W with the slope $\bar{\phi} = 3\pi/4$: initial positions (circles) and paths (solid lines) of the vehicles.

each other. The potential field applied by a neighbor to a vehicle is based on sigmoid functions and designed to attain its minimum at the indicated distance. At each time step, the control law calculates the next move of the vehicle at hand based on minimizing the potential fields applied to this vehicle by all its neighbors. Figure 5.6 shows the resultant paths of the vehicles. Compared to Fig. 5.4, these paths are less regular and the vehicles cover lesser sensing area.

5.5.3   Sweep Coverage along Non-straight Lines

In the real world, the baseline W of the area that should be swept by the vehicles may not be straight; e.g., the coastline of a nation is generally not straight. To test the proposed algorithm in the case of a smooth curved line, we examined a team of ten $n = 10$ autonomous vehicles with the following parameters: $s = 0.3$, $R = 1.5$, $T = 1$, and $v_0 = 0.005$. A typical result of simulation is depicted in Fig. 5.7 and demonstrates a nearly perfect sweep coverage along a smooth curvy line.

Figure 5.8 illustrates the algorithm performance in the case of a broken line, which is a right-angled corner. This scenario may arise if, e.g., a group of security robots should patrol around the perimeter of a building or structure. In the respective simulation, we chose $n = 10$, $r = 1.5$, $s = 0.3$, $R = 2.25$, $T = 1$, and $v_0 = 0.01$. According to Fig. 5.8, the uniform inter-barrier spacing is corrupted for a while when passing the corner, especially for the vehicles that travel closer to the line. However, this

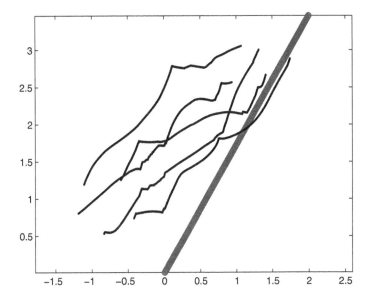

Figure 5.6   Sweep formation using the potential field approach.

spacing is quickly recovered so that the overall behavior is still satisfactory. Thus we see that the straight-line assumption can be relaxed in some situations.

5.5.4   Scalability

Scalability, i.e., capability of adaptation to increasing amount of work and of proper accommodation of added resources on the fly, is an important and desirable property of sensor networks. For example, at some point in time within a large-scale sweep coverage mission, a need may appear to enlarge the coverage area, i.e., to increase the length L of the sensor barrier, with the same detection range $s/2$. Whereas it is inefficacious to interrupt the network operation for accommodation of extra sensors, a better way to address this scalability issue is to let the sensor nodes keep sweeping while introducing new sensor nodes.

In fact, this can be easily achieved by the proposed decentralized motion coordination algorithm: It is only needed to place extra mobile sensors in front of the moving sensor barrier and within its communication range. This is illustrated by Fig. 5.9. A network of initially ten ($n = 10$) mobile sensors with the parameters $r = 1.0$, $s = 0.2$, $R = 1.5$, $T = 1$, and $v_0 = 8 \times 10^{-3}$ encounters the need to increase the monitored margin from the baseline W by 30% in the region $x > 6.5$. To accomplish this increase, three extra sensors are placed at the boundary of this region at $x = 6.5$ and $y = 0.5, 3, 1.7$, respectively. They are activated by communication with the old members of the team. As is shown in Fig. 5.9, three extra sensor nodes join the existing

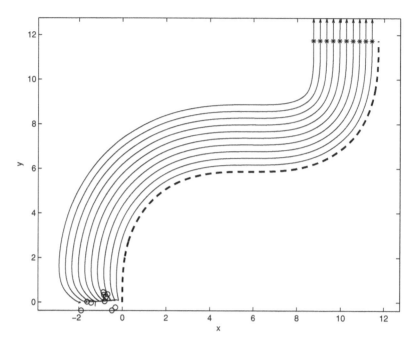

Figure 5.7   Sweep coverage with ten ($n = 10$) autonomous vehicles along a smooth curve: the initial positions (circles) and paths (solid lines) of the vehicles.

network on the fly via the proposed algorithm, extending the length of the sensor barrier and without violation of the deployment density needed for detection.

Scalability property is useful for not only introduction of extra sensors but also replacement of the failed nodes. Figure 5.10 deals with a scenario where a sensor node fails at $x = 4.5$, thus reducing the length of the sensor barrier. To recover the required length, a new sensor was simply added to the network a bit later without interrupting the network operation. In this case, the inequality $r > 2s$ was imposed to maintain the connectivity of the network after the failure of the sensor node. The parameters used for simulation in this case were as follows: $n = 10$, $r = 1.0$, $s = 0.2$, $R = 1.5$, $T = 1$, and $v_0 = 5 \times 10^{-3}$.

5.5.5   Measurement Noises

To assess robustness of the proposed algorithm against uncertainties in the coordination variable ϕ_j and positions p_j of the vehicles, simulation tests were carried out for a network of ten ($n = 10$) vehicles with the following parameters: $r = 1.0$, $s = 0.2$, $R = 1.5$, $T = 1$, and $v_0 = 5 \times 10^{-3}$. The position measurement of any vehicle was corrupted by an additive noise $w_1(kT) \in \mathbb{R}^2$, with its components being random, independent of one another, and uniformly distributed over the interval $[-0.2, 0.2]$.

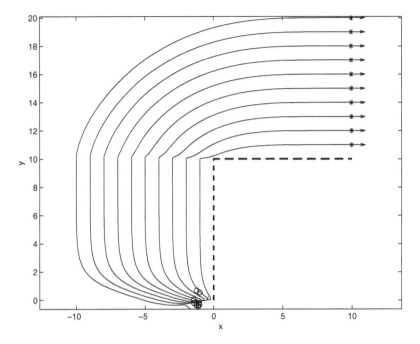

Figure 5.8   Sweep coverage with ten ($n = 10$) autonomous vehicles along a broken line (the right-angled corner): initial positions (circles) and paths (solid lines) of the vehicles.

The coordination variable ϕ_i of every vehicle was also subjected to an additive noise $w_2(kT) \in \mathbb{R}$ uniformly distributed over $[-0.1, 0.1]$.

A typical simulation result is presented in Fig. 5.11(a) and shows that the vehicles can still form an effective sensor barrier with the required deployment density (given by the detection range $s = 0.5$), though their motion expectedly becomes less regular and they do not converge to a perfect straight-line formation. In other words, if the noises are bounded and the sensing and communication ranges are sufficiently large, a moving sensor barrier can still be formed. Figure 5.11(b) illustrates that if subjection to large noises is only an episode in a sweep coverage mission so that the noises vanish $w_1(kT) \to 0, w_2(kT) \to 0$ afterwards, the network recovers a perfect sweep coverage formation, as is described in Definition 5.1.

5.5.6   Sea Exploration

To illustrate a potential application of the proposed algorithm, we carried out a computer simulation that deals with exploration of the Exclusive Economic Zone (EEZ) of the east coast of Australia. This is the region that starts at the coastal baseline and extends 200 nautical miles (370.4 km) out into the sea (see the inset of Fig. 5.12). The objective is to explore the EEZ between 21°S and 37°S. In terms of distance,

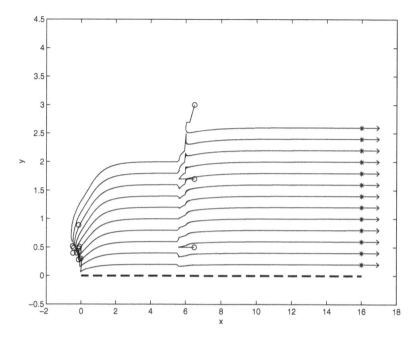

Figure 5.9   Integration of three new team members by a network of initially ten vehicles during a sweep coverage mission.

the length of the concerned coastline is more than 960 nm (1778 km) and hence the region that should be explored is over 6.5×10^5 km^2.

We employed 15 autonomous surface vessels to perform the sea exploration mission. The geographic coordinate system was used, i.e., the position of any vessel was given by its latitude and longitude, termed in degrees. Each vessel was able to communicate with other vessels in the range of 30 nm ($\approx$ 55.6 km, which corresponds to $r = 0.5°$) every 6 minutes (i.e., $T = 0.1$ hours). Also, each vessel was equipped with a sensor that could search the sea floor in the range of 13.2 nm ($\approx$ 24.4 km); in other words, $s = 0.22°$. The desired speed at which the vessels should move along the coastline was set to be 4 knots (i.e., $v_0 = 6.67 \times 10^{-2}$ degrees/hour).

Since the vessels had no prior geographical information about the curvy coast, a land vehicle was deployed that moved over the coastline at the desired speed v_0. The velocity and position of this land vehicle were communicated to the neighboring vessels within the range of 30 nm (which corresponds to 0.5°), thus giving them information about the coastline. At the initial deployment, the sea vessels (i.e., vehicles 1–15) were randomly distributed over the region with longitude 150.32°E–150.32°E and latitude 37.1°S–37.2°S. The initial headings $\theta_i(0)$, $i = 1, 2, \ldots, n$, of the vessels were uniformly distributed over $[0, \pi/2]$.

Figure 5.12 exhibits the respective simulation results for the sea exploration mission. It demonstrates that the vessels form a regular sensor barrier that sweeps along

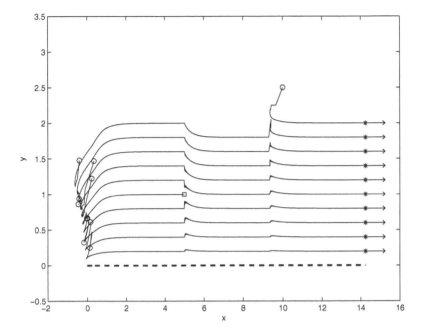

Figure 5.10   A sensor node □ fails during a sweep coverage mission and is replaced by a new node to recover the length of the sensor barrier.

the coastline. The distance from the coastline to the far end of the barrier is approximately $3.3°$. Thus this moving barrier covers a region that starts at the coastal baseline and extends 211nm out into the sea. In other words, the EEZ of the specified coastline can be explored by the autonomous surface vessels using the proposed decentralized control algorithm. In this example, an idealized version of the coast was used, whereas in reality, obstacles, such as islands, may obstruct motions of the vessels along a coastline. This example mainly illustrates that the proposed algorithm has an ability to provide sweep coverage along a boundary that has arbitrary and varying curvatures.

5.6 Proofs of the Technical Facts Underlying Theorem 5.1

Proof of Claim 1 from page 71:

We focus our analysis on $k \geq k_*$ and note first that thanks to (5.11) and (5.27), we have

$$\mathscr{M}_i(kT) := \frac{1}{1 + |\mathscr{N}_i(kT)|} \sum_{j \in \mathscr{N}_i(kT) \cup \{i\}} c_j(kT).$$

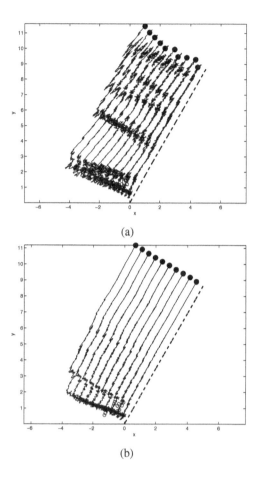

Figure 5.11   Sweep coverage with measurement noises: (a) bounded noise and (b) vanishing noise.

By taking into account (5.2) and (5.18), we have

$$
p_i[(k+1)T] = p_i[kT] + T\hat{v}_i(kT)\begin{bmatrix}\cos\mathcal{H}_i(kT)\\\sin\mathcal{H}_i(kT)\end{bmatrix} + T\bar{v}_i(kT)\begin{bmatrix}\sin\mathcal{H}_i(kT)\\-\cos\mathcal{H}_i(kT)\end{bmatrix}
$$
$$
\overset{(5.26)}{=} p_i[kT] + T\hat{v}_i(kT)\begin{bmatrix}\cos\bar{\phi}\\\sin\bar{\phi}\end{bmatrix} + T\bar{v}_i(kT)\begin{bmatrix}\sin\bar{\phi}\\-\cos\bar{\phi}\end{bmatrix}.
$$

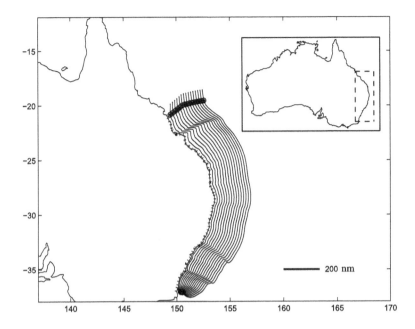

Figure 5.12   Sweep coverage with 15 autonomous vessels along the east coast of Australia: the paths of the vessels (solid lines) and the path of the land vehicle (dots).

So (5.27) (where $k := k+1$) and (5.17) yield that

$$c_i[(k+1)T] = c_i[kT] + [\mathcal{M}_i(kT) - c_i(kT) + v_0 T] = \mathcal{M}_i(kT) + v_0 T$$

$$= \frac{1}{1+|\mathcal{N}_i(kT)|} \sum_{j \in \mathcal{N}_i(kT) \cup \{i\}} c_j(kT) + v_0 T.$$

In terms of the new variables

$$c_i^s(kT) := c_i(kT) - kv_0 T, \tag{5.33}$$

the obtained equation simplifies into

$$c_i^s[(k+1)T] = \frac{1}{1+|\mathcal{N}_i(kT)|} \sum_{j \in \mathcal{N}_i(kT) \cup \{i\}} c_j^s(kT).$$

Thus we see that the variables $c_i^s(kT)$ are updated in accordance with the classic distributed algorithm for reaching consensus [53, Sect. II]. So Assumption 5.3 (in the part concerned with the graph $G(kT)$) entails that the consensus is reached [53]: there exists a constant $\bar{c}$ such that $c_i^s(kT) \to \bar{c}$ as $k \to \infty$ for all $i = 1, \ldots, n$. By (5.33), this is equivalent to (5.28).

Proof of Claim 2 from page 72:

Let us argue via induction on $k = k_*, k_* + 1, \ldots$. For $k = k_*$, inequalities (5.31) hold by construction. Suppose that they are true for some k. Then the first of them is valid for $k := k + 1$ as well, as was shown on page 72. It remains to show that

$$q_i[(k+1)T] < q_{i+1}[(k+1)T] \tag{5.34}$$

for $i = 1, \ldots, n-1$. We shall consider separately the cases where $i = 1$ and $1 < i < n$, respectively, and employ the fact that due to (5.30) and the square shape of the communication zone, vehicles $j \neq j'$ are neighbors at time kT if and only if

$$\left| q_{j'}(kT) - q_j(kT) \right| \leq r.$$

(A) $i = 1$. Here we also consider the feasible cases separately.

- *Vehicles* 1 *and* 2 *are not neighbors:* $q_2(kT) - q_1(kT) > r$.

  - *Vehicle* 2 *detects the baseline W:* $q_2(kT) \leq R$. Then so does vehicle 1 as well. Hence (5.16) implies that $q_1[(k+1)T] = 1/2[q_1(kT) + s/2]$, whereas $q_2[(k+1)T] = 1/2[q_2(kT) + s/2]$ if vehicle 2 has no neighbors $q_j(kT)$ to the right; otherwise, $q_2[(k+1)T] \geq 2/3 q_2(kT)$. In the first case, (5.34) is clearly true. In the "otherwise" case,

$$q_2[(k+1)T] - q_1[(k+1)T] \geq \frac{2}{3} q_2(kT) - \frac{1}{2}\left[q_1(kT) + \frac{s}{2}\right]$$
$$\geq \frac{2}{3} q_2(kT) - \frac{1}{2}[q_2(kT) - r + s] = \frac{r-s}{2} + \frac{1}{6} q_2(kT) > 0,$$

where the last inequality is valid thanks to (5.19). Thus (5.34) does hold.

  - *Vehicle* 2 *does not detect the baseline W.* Depending on whether sensor 1 detects the baseline or not, either $q_1[(k+1)T] = 1/2[q_1(kT) + s/2]$ or $q_1[(k+1)T] = q_1[kT]$ by (5.16); in any case, $q_1[(k+1)T] \leq 1/2[q_1(kT) + s/2]$. Similarly, $q_2[(k+1)T] \geq q_2(kT) - s/2$ and so

$$q_2[(k+1)T] - q_1[(k+1)T] \geq q_2(kT) - \frac{s}{2} - \frac{1}{2}[q_1(kT) + s/2]$$
$$\geq q_2(kT) - q_1(kT) - \frac{3}{4}s \geq r - \frac{3}{4}s > 0,$$

where the last inequality is valid thanks to (5.19). Thus (5.34) does hold.

- *Vehicles* 1 *and* 2 *are neighbors.*

  - *Vehicle* 1 *detects the baseline W.* Then $q_1[(k+1)T] = 1/2q_1[kT] + 1/6q_2[kT]$; whereas $q_2[(k+1)T] = 1/2[q_1(kT) + q_2(kT) + s]$ if vehicle 2 has no neighbors to the right, and $q_2[(k+1)T] \geq 1/3q_1(kT) + 2/3q_2(kT)$ otherwise. In both cases, (5.34) clearly does hold.

- *Vehicle 1 does not detect the baseline W.* Then $q_1[(k+1)T] = 1/2[q_1(kT) + q_2(kT) - s]$. Hence (5.34) clearly is clearly true if vehicle 2 has no neighbors to the right. Otherwise

$$q_2[(k+1)T] - q_1[(k+1)T] \geq \frac{1}{3}q_1(kT) + \frac{2}{3}q_2(kT)$$
$$- \frac{1}{2}[q_1(kT) + q_2(kT) - s] = \frac{1}{6}[q_2(kT) - q_1(kT)] + \frac{s}{2} > 0,$$

which completes the proof of (5.34).

(B) $1 < i < n$. Again, we consider several feasible cases separately.

- *Vehicles i and $i+1$ are not neighbors:* $q_{i+1}(kT) - q_i(kT) > r$. Then $q_i[(k+1)T] < q_i(kT) + \frac{s}{2}$ and $q_{i+1}[(k+1)T] > q_{i+1}(kT) - \frac{s}{2}$ by (5.16). Thus

$$q_{i+1}[(k+1)T] - q_i[(k+1)T] > q_{i+1}(kT) - \frac{s}{2} - q_i(kT) - \frac{s}{2} > r - s > 0,$$

where the last inequality is valid thanks to (5.19). Thus (5.34) does hold.

- *Vehicles i and $i+1$ are neighbors.*

  - *Vehicle i has no neighbors to the left and does not detect the baseline.* Then $q_i[(k+1)T] = 1/2[q_i(kT) + q_{i+1}(kT) - s]$. Meanwhile $q_{i+1}[(k+1)T] = 1/2[q_i(kT) + q_{i+1}(kT) + s]$ if vehicle $i+1$ has no neighbors to the right; otherwise, $q_{i+1}[(k+1)T] > 1/3q_i(kT) + 2/3q_{i+1}(kT)$. In the first case, (5.34) is evident. In the second case, (5.34) holds since

$$q_{i+1}[(k+1)T] - q_i[(k+1)T] > \frac{1}{3}q_i(kT) + \frac{2}{3}q_{i+1}(kT)$$
$$- \frac{1}{2}[q_i(kT) + q_{i+1}(kT) - s] = \frac{1}{6}[q_{i+1}(kT) - q_i(kT)] + \frac{s}{2} > 0.$$

  - *Vehicle i either has a neighbor to the left or detects the baseline.* Then $q_i[(k+1)T] < 2/3q_i(kT) + 1/3q_{i+1}(kT)$. As before, $q_{i+1}[(k+1)T] = 1/2[q_i(kT) + q_{i+1}(kT) + s]$ if vehicle $i+1$ has no neighbors to the right; otherwise, $q_{i+1}[(k+1)T] > 1/3q_i(kT) + 2/3q_{i+1}(kT)$. In the first case,

$$q_{i+1}[(k+1)T] - q_i[(k+1)T] > \frac{1}{2}[q_i(kT) + q_{i+1}(kT) + s]$$
$$- \frac{2}{3}q_i(kT) - \frac{1}{3}q_{i+1}(kT) = \frac{1}{6}[q_{i+1}(kT) - q_i(kT)] + \frac{s}{2} > 0.$$

In the second case,

$$q_{i+1}[(k+1)T] - q_i[(k+1)T] > \frac{1}{3}q_i(kT) + \frac{2}{3}q_{i+1}(kT)$$
$$- \frac{2}{3}q_i(kT) - \frac{1}{3}q_{i+1}(kT) = \frac{1}{3}[q_{i+1}(kT) - q_i(kT)] + \frac{s}{2} > 0.$$

Thus we see that (5.34) is valid in both cases.

Proof of Claim 3 from page 72:

We still focus our analysis on $k \geq k_*$. Since then the square communication zone of any vehicle is aligned with the baseline, Claim 2 and (5.30) imply that the following statements hold for any $i = 1, \ldots, n$ and the indices α, β from (5.15):

- Index α may take only one value $i - 1$ and is treated as existing if and only if

$$
\begin{aligned}
q_i(kT) - q_{i-1}(kT) &\leq r \quad \text{for} \quad i \geq 2, \\
q_i(kT) - q_{i-1}(kT) &\leq R \quad \text{for} \quad i = 1;
\end{aligned}
\tag{5.35}
$$

- Index β may take only one value $i + 1$ and is treated as existing if and only if

$$
i \leq n - 1 \quad \text{and} \quad q_{i+1}(kT) - q_i(kT) \leq r.
\tag{5.36}
$$

So it is easy to see by inspection that putting $q_j^i(kT) := \mathfrak{h}_j(kT), j = 0, \ldots, n$, transforms the the right-hand side of (5.16) into $\mathfrak{h}_i((k+1)T)$ for all $i = 1, \ldots, n$. Since the original right-hand side equals $q_i((k+1)T)$, it follows that the discrepancy $\partial_i(kT) := q_i(kT) - \mathfrak{h}_i(kT)$ is updated in accordance with the following rule for any $i = 1, \ldots, n$:

$$
\partial_i((k+1)T) =
\begin{cases}
\frac{1}{3}[\partial_{i-1}(kT) + \partial_i(kT) + \partial_{i+1}(kT)] & \text{if both (5.35) and (5.36) hold, and } i \geq 2; \\
\frac{1}{3}\partial_{i-1}(kT) + \frac{1}{2}\partial_i(kT) + \frac{1}{6}\partial_{i+1}(kT) & \text{if both (5.35) and (5.36) hold, and } i = 1; \\
\frac{1}{2}[\partial_{i-1}(kT) + \partial_i(kT)] & \text{if only (5.35) holds;} \\
\frac{1}{2}[\partial_{i+1}(kT) + \partial_i(kT)] & \text{if only (5.36) holds;} \\
\partial_i(kT) & \text{if both (5.35) and (5.36) do not hold.}
\end{cases}
$$

Meanwhile $\partial_0(kT) \equiv 0, k \geq k_*$.

Now we introduce a set of artificial "agents," labelled 0 through n, where agent 0 is treated as a team leader and any agent i is endowed with the scalar coordination variable $\partial_i(kT)$. The above rule of updating this variable is in fact a special case of the classic distributed algorithm for reaching consensus in a team with a leader [5, 53, 92, 119]. In general, this algorithm is expressed by a set of linear equations

$$
\partial_i((k+1)T) = \sum_{j=0}^{n} a_{ij}(k)\partial_j(kT), \qquad i = 1, \ldots,
$$

whose coefficients satisfy the following properties with some $a > 0$ (which is independent of i and k):

- For any k, any coefficient $a_{ij}(k)$ is either zero or no less that a;
- $a_{ii}(k) \neq 0$ for all $i = 0, \ldots, n$ and k;
- $\sum_{j=0}^{n} a_{ij}(k) = 1$ for all $i = 1, \ldots, n$ and k.

In the case at hand, these properties are evidently satisfied with $a = 1/6$ since $a_{00}(k) \equiv 1$ and the other nonzero coefficients are as follows:

$$\begin{cases} a_{i-1,i}(k) = a_{ii}(k) = a_{i,i+1}(k) = \frac{1}{3} & \text{if (5.35) and (5.36) hold, and } i \geq 2; \\ a_{i-1,i}(k) = \frac{1}{3}, a_{ii}(k) = \frac{1}{2}, a_{i,i+1}(k) = \frac{1}{6} & \text{if (5.35) and (5.36) hold, and } i = 1; \\ a_{i-1,i}(k) = a_{ii}(k) = \frac{1}{2} & \text{if only (5.35) holds;} \\ a_{ii}(k) = a_{i,i+1}(k) = \frac{1}{2} & \text{if only (5.36) holds;} \\ a_{ii}(k) = 1 & \text{if both (5.35) and (5.36) do not hold.} \end{cases}$$

To apply the results of [92], we introduce a directed graph $\widetilde{G}(kT)$ for any $k \geq k_*$. Its vertices are identified with the integers $i = 0, \ldots, n$, and there is an edge from i to $j \neq i$ if and only if $a_{ij}(k) > 0$. It should be shown that the so introduced sequence of graphs satisfies a modified Main Connectivity Assumption, i.e., Assumption 2.4 from Chapter 2 stated on page 11, where the request for connectivity of the graph is replaced by the requirement that it contains a directed spanning tree. We shall seek a tree rooted at the leader 0, and start with two preliminary observations. First, the graph that results from $\widetilde{G}(kT)$ by dropping the leader 0 with outcoming edges is in fact undirected: If there is an edge from i to j, there also is an edge from j to i. By welding these converse directed edges into a single undirected edge, we obtain an undirected graph $\widehat{G}(kT)$. Second, since the standard Main Connectivity Assumption (Assumption 2.4 from Chapter 2 stated on page 11) holds for $\bar{G}(kT)$ by Assumption 5.3, any of the intervals $[k_j : k_{j+1})$ from Assumption 2.4 contains k such that the graph $\bar{G}(kT)$ links 0 with a vertex $i \geq 1$: Vehicle i detects the baseline at time $t = kT$. For $k \geq k_*$, (5.31) implies that then so does vehicle 1, and hence $\widetilde{G}(kT)$ contains an edge from the prospective root 0 to 1. These preliminaries show that it suffices to check the standard Main Connectivity Assumption for $\widehat{G}(kT)$.

To this end, we note that edges of $\widehat{G}(kT)$ may connect only vertices of the form i and $i + 1$, where $i = 1, \ldots, n - 1$; and these two vertices are connected if and only if

$$|q_i(kT) - q_{i+1}(kT)| \leq r \text{ and } i \geq 1. \tag{5.37}$$

Now we invoke (5.30) and (5.31) and that the square communication zone of any vehicle is aligned with the baseline for $k \geq k_*$. So (5.37) means that vehicles i and $i + 1$ are neighbors. It follows that $\widehat{G}(kT)$ is a subgraph of the graph $G(kT)$ addressed by Assumption 5.3: These graphs have a common set of vertices, whereas the edges of $\widehat{G}(kT)$ form a subset of edges of graph $G(kT)$. The converse may be incorrect, but the following similar, though slighter, property holds for $k \geq k_*$:

(P) If nodes $i \neq j$ are connected with an edge in graph $G(kT)$, they are connected with a path in $\widehat{G}(kT)$.

Indeed vehicles i and j are neighbors at time $t = kT$ by the definition of $G(kT)$. So $|q_i(kT) - q_j(kT)| \leq r$ and due to (5.31), $|q_l(kT) - q_{l+1}(kT)| \leq r$ for any $l = \min\{i, j\}, \ldots, \max\{i, j\} - 1$, i.e., l and $l + 1$ are connected with an edge in $\widehat{G}(kT)$ for any such l. Hence the conclusion of P) does hold.

It follows that the sequence of undirected graphs $\widehat{G}(kT), k = k_*, k_* + 1, \ldots$ does satisfy the standard Main Connectivity Assumption, and so the sequence of directed graphs $\widetilde{G}(kT), k = k_*, k_* + 1, \ldots$ satisfies its modified version described above. This implies that as $k \to \infty$, the coordination variables $\eth_i(kT)$ of all agents $i = 1, \ldots, n$ converge to the value 0 constantly assumed by the coordination variable of the team leader 0 [92, Theorem 2.39]. The proof of (5.32) is completed by invoking that $\eth_i(kT) = q_i(kT) - \hbar_i(kT)$.

CHAPTER 6

OPTIMAL DISTRIBUTED BLANKET COVERAGE PROBLEM

6.1 Introduction

In Chapter 1, three types of coverage problems for sensor networks were introduced. They are blanket coverage, barrier coverage, and sweep coverage. This chapter addresses the blanket coverage problem which is the most difficult one. The complete blanket coverage problem studied in this chapter is to deploy a network of mobile robotic sensors so that every point of a given planar region is sensed by at least one of the sensors. The region for deployment is bounded and of an arbitrary shape, which is not known to the mobile robotic sensors a priori. We propose a decentralized randomized control algorithm that drives the network of mobile sensors to form a grid consisting of equal equilateral triangles. As always in this book, the proposed algorithm is based on simple local control rules that are based only on information about the closest neighbors of each sensor. A major benefit of deploying sensors in the triangular lattice pattern is that it is asymptotically optimal in terms of minimum number of sensors required for the complete coverage of an arbitrary bounded set. This fact follows from the important mathematical result of Kershner [56].

The triangular lattice pattern for mobile sensor networks was studied in [13]; however, the heuristic algorithm of [13] was only verified by simulation studies and no

Decentralized Coverage Control Problems for Mobile Robotic Sensor and Actuator Networks.
By Andrey V. Savkin, Teddy M. Cheng, Zhiyu Xi, Faizan Javed, Alexey S. Matveev, and Hung
Nguyen. Copyright © 2015 by the Institute of Electrical and Electronics Engineers, Inc.

theoretically rigorous analysis was given. Moreover, in [13], the boundaries of the monitored region have to be known by all the sensors a priori, which is a severe constraint in practice. Unlike many coverage algorithms proposed in this area, our algorithm is theoretically verified. In particular, we give a mathematically rigorous proof of convergence of our algorithm with probability 1 for any initial positions of the mobile sensors. There are many papers on blanket coverage control in mobile networks based on Voronoi partitions techniques; see, e.g., [7] and references therein. However, the Voronoi partitions approach requires solving some quite difficult computational geometry problems, whereas the blanket coverage control of this chapter is computationally highly efficient, which is very attractive especially for mobile robotic networks with limited computing powers.

The main results of this chapter are originally published in [101].

Various modifications of the randomized algorithm of this chapter are successfully applied to totally different problems such as formation building for non-holonomic robots [109] and avoiding cascaded failures in complex networks [2] and power networks [3]. Furthermore, three-dimensional versions of the results of this chapter are presented in [83].

The remainder of the chapter is organized as follows. Section 6.2 formulates the blanket coverage problem studied in this chapter. Section 6.3 presents a randomized algorithm of distributed self-deployment for blanket coverage and its mathematical analysis. Finally, Section 6.4 contains computer simulation results illustrating the proposed algorithm.

6.2 Blanket Coverage Problem Formulation

Our objective is to develop a distributed algorithm to drive a network of autonomous mobile sensors to completely cover a bounded two-dimensional region $\mathscr{R} \subset \mathbb{R}^2$. We do not assume that $\mathscr{R}$ is simply connected, hence, $\mathscr{R}$ can have "holes" (see Fig. 6.1). We consider a network consisting of n mobile sensors, labelled 1 through n, and denote by $p_i \in \mathbb{R}^2$ the vector of the Cartesian coordinates of sensor i. Each mobile sensor has a sensing range of $r_s > 0$ and communicates with its surrounding neighbors in a range of $r_c \geq \sqrt{3}r_s$ at the discrete time instances $k = 0, 1, 2, \ldots$ for the coordination of their motions. In other words, at time k sensor i has the ability to obtain information from its neighbors in a disk of radius r_c defined by $D_{i,r_c}(k) := \{p \in \mathbb{R}^2 : \|p - p_i(k)\| \leq r_c\}$, where $\|\cdot\|$ is the Euclidean norm. The region $\mathscr{R}$ is unknown to the sensors a priori. Let $\mathscr{R}(r_s)$ denote the r_s-vicinity of the set $\mathscr{R}$, i.e. $\mathscr{R}(r_s)$ consists of all points that either belong to $\mathscr{R}$ or are at a distance of r_s or less from $\mathscr{R}$. In terms of detection, each sensor can detect or sense any object in a disk of radius r_s. Furthermore, each sensor can detect the boundary of $\mathscr{R}$ in a range of r_c. Moreover for any point p in a range of r_c, the sensor can determine whether this point belongs to $\mathscr{R}(r_s)$. The mobile sensors are allowed to cross the boundary of $\mathscr{R}$.

Only a minor assumption is imposed on $\mathscr{R}$.

Assumption 6.1 *The region $\mathscr{R}$ is bounded, connected, and Lebesgue measurable.*

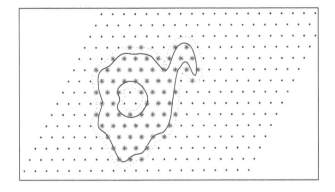

Figure 6.1   Bounded two-dimensional region; its r_s-vicinity is bounded by the dotted lines, the points of $\mathcal{V}$ are denoted by '·', and the points of $\hat{\mathcal{V}}$ are denoted by '*'.

Definition 6.1 *A finite set of points $\hat{\mathcal{W}}$ is said to be a complete blanket coverage of the region $\mathcal{R}$ if for any $p \in \mathcal{R}$ there exists a point $w \in \hat{\mathcal{W}}$ such that $\|p - w\| \le r_s$.*

It should be noted that some points of $\hat{\mathcal{W}}$ may be outside $\mathcal{R}$.

Definition 6.2 *Consider a triangular grid cutting the plane into equilateral triangles with the sides of length $\sqrt{3}r_s$. Let $\mathcal{V}$ be the infinite set of all vertices of this grid. The set $\hat{\mathcal{V}} := \mathcal{V} \cap \mathcal{R}(r_s)$ is called a* triangular covering set *of $\mathcal{R}$ (see Fig. 6.1).*

Any triangular covering set of $\mathcal{R}$ is a complete blanket coverage of $\mathcal{R}$. Indeed, for any $p \in \mathcal{R}$ there evidently exists $v \in \mathcal{V}$ such that $\|p - v\| \le r_s$ (v is the vertex of the triangular grid that is closest to p). Hence, $v \in \mathcal{R}(r_s)$ and so $v \in \hat{\mathcal{V}}$.

Thus, if the sensor network is deployed so that there is a sensor at each point of some triangular covering set, a complete blanket coverage of the region $\mathcal{R}$ is achieved. Moreover, the following theorem, which can be easily derived from the main result of Kershner [56], shows that triangular coverage is asymptotically optimal: For small r_s, it contains almost the minimum possible number of elements.

Theorem 6.1 *Let Assumption 6.1 hold and let $\mathbf{ar}(\mathcal{R})$ denote the Lebesgue measure or area of $\mathcal{R}$. For any $r_s > 0$, let $\hat{\mathcal{V}}(r_s)$ be a triangular covering set of $\mathcal{R}$, and let $t(r_s)$ stand for the number of points in $\hat{\mathcal{V}}(r_s)$. Also, let $m(r_s)$ be the minimum possible number of points, where the minimum is over all complete blanket coverages of $\mathcal{R}$. Then $\lim_{r_s \to 0} r_s^2 t(r_s) = \lim_{r_s \to 0} r_s^2 m(r_s) = \frac{2\sqrt{3}}{9} \mathbf{ar}(\mathcal{R})$.*

Our aim is to develop a distributed algorithm that ensures deployment of the sensors at all vertices of a triangular covering set.

6.3 Randomized Coverage Algorithm

For any angle θ, we introduce the vectors $n_1(\theta) := [\cos(\theta), \sin(\theta)]^T$ and $n_2(\theta) := [\cos(\theta + \frac{\pi}{3}), \sin(\theta + \frac{\pi}{3})]^T$. These vectors together with the vector $n_1(\theta) - n_2(\theta)$ determine the headings of three lines of a triangular grid. Any triangular grid is uniquely defined by an angle θ and any of its vertices q. Hence any θ and q uniquely determine a triangular covering set of $\mathcal{R}$, which will be denoted by $\hat{\mathcal{V}}[q, \theta]$.

Let each sensor i form its own opinion about the grid to be used. At time k this opinion is expressed by the values of two *consensus variables* $\theta_i(k) \in \mathbb{R}$ and $q_i(k) \in \mathbb{R}^2$, where $\theta_i(k)$ and $q_i(k)$ give the orientation of a triangular grid and its vertex, respectively. The first idea behind the proposed algorithm is that though the sensors may and typically do start with different values of $\theta_i(\cdot)$ and $q_i(\cdot)$, these variables should be constantly updated in a way that ensures their convergence to some consensus values θ_0 and q_0: these values are common for all sensors and so define a common triangular grid. It is this grid that in fact will be used to form a triangular coverage.

To describe the proposed update rules, we need some definitions and notations. Sensor j is called the *neighbor* of sensor i at time $t = kT$ if at this time it is within the communication range of sensor i, i.e., lies in the disk $D_{i, r_c}(k)$. Thus neighborship means ability to communicate with each other. Let $\mathcal{N}_i(k)$ stand for the set of all neighbors of sensor i at time $t = kT$, and let $|\mathcal{N}_i(k)|$ be the number of elements in $\mathcal{N}_i(k)$ (0 if $\mathcal{N}_i(k)$ is empty). Finally for an angle θ and two points q and p on the plane, we denote by $C[q, \theta](p)$ the vertex of the triangular covering set $\hat{\mathcal{V}}[q, \theta]$ that is closest to p (if there are several such vertices, we take any of them).

We propose the following rules for updating the consensus variables $\theta_i(k)$, $q_i(k)$ and the sensors' coordinates $p_i(k)$:

$$\theta_i(k+1) = \frac{\theta_i(k) + \sum_{j \in \mathcal{N}_i(k)} \theta_j(k)}{1 + |\mathcal{N}_i(k)|};$$

$$q_i(k+1) = \frac{q_i(k) + \sum_{j \in \mathcal{N}_i(k)} q_j(k)}{1 + |\mathcal{N}_i(k)|};$$

$$(6.1)$$

$$p_i(k+1) = C[q_i(k), \theta_i(k)](p_i(k)). \qquad (6.2)$$

The second rule (6.2) drives the sensor to the nearest vertex of the triangular covering set that is formed according to the current opinion of this sensor about the orientation and location of the underlying triangular grid. The local rule (6.1) merely averages the opinions of the neighbors, including the sensor at hand, about the triangular grid heading and phase shift, with an ultimate aim to achieve consensus on the grid among all team members. Such a rule was used in numerous papers on control of multi-agent systems (see, e.g., [10, 53]) and, under certain technical assumptions, does guarantee consensus: All consensus variables converge to values that are common for all agents [53]. Among those assumptions, an analog of the standard Main Connectivity Assumption (Assumption 2.4 from Chapter 2) typically plays a key role.

To state it in the current context, we visualize the relation of neighborship: At any time k, we introduce an undirected graph $G(k)$ with the vertices $1, 2, \ldots, n$ that links vertices $i \neq j$ with an undirected edge if and only if sensors i and j are neighbors at time k. In terms of this graph, we impose the following.

Assumption 6.2 *There exists an infinite sequence of contiguous, non-empty, bounded discrete time-intervals $[k_j : k_{j+1})$, $j = 0, 1, 2, \ldots$, starting at $k_0 = 0$, such that the union of $G(k)$ over any interval $[k_j, k_{j+1})$ is a connected graph.*

Since the covering set $\hat{\mathscr{V}}[q, \theta]$ is invariant with respect to rotation through an angle of $\pi/3$ around any of its vertices $\hat{\mathscr{V}}[q, \theta] = \hat{\mathscr{V}}[q, \theta + \pi/3]$, the initial values $\theta_i(0)$ can be chosen from any prespecified interval $[\theta_-, \theta_+)$ of length $\theta_+ - \theta_- \geq \pi/3$ without any loss of generality.

Theorem 6.2 *Let Assumptions 6.1 and 6.2 hold and the mobile sensors move in accordance with the decentralized control algorithm (6.1), (6.2). Then there exists a triangular covering set $\hat{\mathscr{V}}$ such that any sensor $i = 1, 2, \ldots, n$ converges to this set:*

$$\mathbf{dist}\left[p_i(k), \hat{\mathscr{V}}\right] := \min_{v \in \hat{\mathscr{V}}} \|p_i(k) - v\| \to 0 \quad as \quad k \to \infty. \tag{6.3}$$

Furthermore, the sensors come to a consensus on the grid: there exist $\theta_0 \in \mathbb{R}$ and $q_0 \in \mathbb{R}^2$ such that

$$\theta_i(k) \to \theta_0 \quad and \quad q_i(k) \to q_0 \quad for \ all \quad i = 1, 2, \ldots, n. \tag{6.4}$$

Proof: Assumption 6.2 and the update rule (6.1) guarantee that (6.4) does hold [53]. Now we take $\hat{\mathscr{V}} := \hat{\mathscr{V}}[q_0, \theta_0]$. Then (6.3) is straightforward from the update rule (6.2) and (6.4). ∎

Theorem 6.2 guarantees only that the algorithm (6.1), (6.2) drives all the sensors to vertices of some triangular covering set, but it does not guarantee that all vertices will be occupied by the sensors. However, the last property is crucial to achieve complete coverage of the area. Therefore, we develop the second stage of the algorithm. Now we assume that all the mobile sensors are at vertices of a triangular covering set and move from one vertex to another at discrete times, i.e., $p_i(k) \in \hat{\mathscr{V}}$ for all i and k. The objective is to occupy all the vertices of this set.

To state the respective algorithm, we denote by $\mathscr{S}[p_i(k)]$ the set that consists of the following vertices of the triangular covering set $\hat{\mathscr{V}}$:

- The vertex $p_i(k)$ currently occupied by sensor i;

- The neighboring vertices $v \in \hat{\mathscr{V}}$ that are vacant at time $t = kT$, i.e., such that $v \neq p_j(kT)$ for all $j = 1, \ldots, n$.

Here "neighborship" means "neighborship in the triangular grid" so that any vertex $v \in \hat{\mathscr{V}}$ has six neighboring vertices in the grid. However, not all of them may lie in the associated triangular covering set $\hat{\mathscr{V}}$, which is the intersection of the grid with the r_s-vicinity of the region $\mathscr{R}$. So v has at most six neighbors in the covering set. Hence the number $|\mathscr{S}[p_i(k)]|$ of elements in $\mathscr{S}[p_i(k)]$ obeys the inequalities $1 \leq |\mathscr{S}[p_i(k)]| \leq 7$.

The following randomized algorithm is applied to any sensor $i = 1, \ldots, n$:

- With probability $\frac{1}{|\mathscr{S}[p_i(k)]|}$, a vertex $v \in \mathscr{S}[p_i(k)]$ is selected; and then the sensor is moved to this vertex:

$$p_i(k+1) = v \quad \text{with probability} \quad \frac{1}{|\mathscr{S}[p_i(k)]|} \quad \forall v \in \mathscr{S}[p_i(k)]. \quad (6.5)$$

It is assumed that selections made by different sensors, as well as selections made by the same sensor at different times, are independent of one another.

The rule (6.5) evidently implies that no sensor moves whenever all the vertices of the triangular covering set are occupied. The last and desirable situation can be achieved only if there are sufficiently many sensors, as is specified in the following.

Assumption 6.3 *The number n of the sensors is no less than the maximum possible number t_M of vertices in the triangular covering set.*

By Theorem 6.1, $t_M \approx \frac{2\sqrt{3}}{9r_s^2}\mathbf{ar}(\mathscr{R})$ for small r_s.

Theorem 6.3 *Suppose that Assumption 6.3 holds and the mobile sensors move in accordance with the decentralized control algorithm (6.5). Then with probability 1, there exists a time $k_* \geq 0$ such that all vertices of the triangular covering set $\hat{\mathscr{V}}$ are occupied by sensors at any time instant $k \geq k_*$.*

Proof: At any time, the state of the network can be identified with a map $f(\cdot)$ from the set of the sensors $\{1, \ldots, n\}$ to the finite covering set $\hat{\mathscr{V}}$: Every sensor i is mapped into the vertex that currently accommodates this sensor. The variety $\mathfrak{F}$ of all maps $f(\cdot) : \{1, \ldots, n\} \to \hat{\mathscr{V}}$ is finite. The state of the network evolves over this set in accordance with the randomized rule (6.5), thus forming a random process $f_1, f_2, \ldots \in \mathfrak{F}$. It is easy to see that, firstly, this process has the Markov property (given that the present state, the future, and past states are independent) and, secondly, this Markov chain is time-homogeneous. With probability 1, the system cannot leave any state f for which all vertices of $\hat{\mathscr{V}}$ are occupied, as was remarked. In other words, such a state is absorbing [44, Ch. 11]. Thanks to Assumptions 6.1 and 6.3, these absorbing states do exist and exhaust all absorbing states. Moreover, it is easy to see that from any initial state, the system may go to an absorbing state with a nonzero probability, not necessarily in one step. This means that the Markov chain itself is absorbing [44, Ch. 11], and so implies that with probability 1, one of the absorbing states is inevitably reached [44, Ch. 11]. This completes the proof of Theorem 6.3. ∎

After all the vertices of the triangular covering set $\hat{\mathscr{V}}$ are occupied by the sensors, some vertices must simultaneously host multiple sensors if the total number of sensors exceeds the number of vertices in $\hat{\mathscr{V}}$. In this chapter, we do not address the issue of avoiding collisions between mobile sensors by assuming that whenever several sensors intend to move in a common vertex, collisions among them are resolved via their placement at different locations in a close proximity of this vertex.

As was shown in the proof of Theorem 6.3, desirable deployments of the sensor network (where all vertices of the triangular covering set $\mathcal{V}$ are occupied) correspond to absorbing states of a discrete-time finite Markov chain. So the mean time T_d of achieving such a deployment can be evaluated via standard techniques of computing the mean time of reaching an absorbing state in a Markov chain; see, e.g., [44, Ch. 11]. However, for large-scale networks, this approach is typically computationally expensive, involving work with large size matrices, and anyhow this time is typically large. Development of alternative and more effective methods for estimation of T_d, as well as modifications of the proposed algorithm in order to reduce T_d, are topics of our ongoing research.

6.4   Illustrative Examples

In the first example, we consider a region $\mathcal{R}$ from Fig. 6.2(a), along with 80 sensors whose positions are displayed by $\cdot$ in Fig. 6.2(a). Initially the sensors are randomly deployed at the upper left side of $\mathcal{R}$, as is shown in Fig. 6.2(a). The initial value of the consensus variable $\theta_i(0)$ of sensor i is randomly and uniformly distributed over the range of $\left(0, \frac{7}{18}\pi\right)$. The initial value of the consensus variable $q_i(k)$ is randomly and uniformly distributed over a square with a side length of $8.5r_s$. Driven by the algorithm (6.1), (6.2), the mobile sensors come in vertices of an agreed triangular grid at time $k = 35$, as is shown in Fig. 6.2(b) where the sensors' positions are denoted by $\circ$. After this the algorithm (6.5) was applied to obtain a complete blanket coverage. Figure 6.3(a) shows the sensors' locations after 12 steps of this algorithm; these locations already cover a larger part of the region but not the entire region. The complete coverage is achieved at $k = 22$, as is shown in Fig. 6.3(b): All the vertices of the triangular covering set are occupied, and the region $\mathcal{R}$ is completely covered by the sensors. The same experiment was repeatedly run with different initial conditions selected according to the aforementioned distributions. In the course of this series, the time of reaching a complete coverage deployment varied between 10 and 22 iterations.

The second example is concerned with the region $\mathcal{R}$ from Fig. 6.4(a) and 100 sensors initially deployed randomly at the lower right side of $\mathcal{R}$, as is shown in Fig. 6.4(a). The initial value of the consensus variable $\theta_i(0)$ of sensor i is randomly and uniformly distributed over the range of $\left(0, \frac{5}{18}\pi\right)$. The initial value of the consensus variable $q_i(k)$ is randomly and uniformly distributed over a square with a side length of $8.5r_s$. Driven by the algorithm (6.1), (6.2), the sensors come in vertices of an agreed triangular grid at time $k = 40$, as is shown in Fig. 6.4(b) where the sensors' positions are denoted by $\circ$. After this the algorithm (6.5) was applied to obtain a complete blanket coverage. Figure 6.4(c) shows that after 5 steps, the sensors do not cover the entire region. A complete coverage is achieved after 26 steps, as is shown in Fig. 6.4(d). The same experiment was repeatedly run with different initial conditions selected according to the aforementioned distributions. In the course of this series, the mean time of reaching a complete coverage deployment was approximately 11 iterations.

The simulations were carried out using MATLAB R2009a (The Mathworks Inc, Natick, MA) on an Intel Core 2 DUO 2.53 GHz processor with 4 GB RAM running Windows 7. Of course, the number of iterations might increase for regions $\mathcal{R}$ with very non-convex and complex shapes.

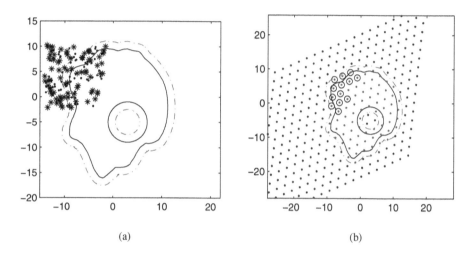

(a)   (b)

Figure 6.2   (a) Initial deployment. (b) Outcome of the algorithm (6.1), (6.2) at $k = 35$.

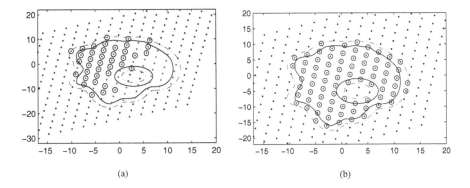

Figure 6.3 Outcome of the control law (6.5) at (a) $k = 12$ and (b) $k = 22$.

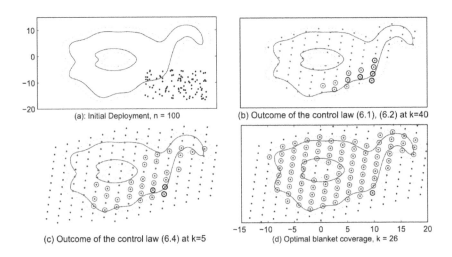

Figure 6.4 Movement of hundred sensors to a complete blanket coverage: (a) Initial deployment; (b) Consensus on a triangular grid at $k = 40$; (c) Incomplete coverage at $k = 5$; (d) Blanket coverage formed by the sensors at $k = 26$.

CHAPTER 7

DISTRIBUTED SELF-DEPLOYMENT FOR FORMING A DESIRED GEOMETRIC SHAPE

7.1 Introduction

In this chapter, we consider a case of the so-called flocking problem for a group of mobile robots. In flocking problems, the control objective is to cooperatively drive mobile robots from random initial locations to a formation of a desired geometric shape. The cooperation takes place in the form of coordinated control among the robotic sensors using the information from the network. However, to reduce the cost of an operation, each robotic sensor may have very limited resources, e.g., communication or sensing powers. Therefore, a centralized control algorithm is not a practical approach since not all the global information is available to each sensor. To circumvent the lack of global information, a decentralized approach should be considered.

The study of decentralized control laws for groups of autonomous sensors has emerged as a challenging research area recently; see, e.g., [45, 53, 69, 95, 96, 106, 131, 134]. In this control framework, the motion of each sensor is coordinated using local information such as coordinates or velocities of several other sensors that are currently the closest neighbors of the sensor at hand. One approach to developing these local motions is based on the so-called consensus or agreement scheme, as in

Decentralized Coverage Control Problems for Mobile Robotic Sensor and Actuator Networks.
By Andrey V. Savkin, Teddy M. Cheng, Zhiyu Xi, Faizan Javed, Alexey S. Matveev, and Hung Nguyen. Copyright © 2015 by the Institute of Electrical and Electronics Engineers, Inc.

coverage control problems studied in the previous chapters of this book. By using this scheme, a network of mobile sensors can coordinate to achieve, e.g., a specific formation or geometric structure. One of the advantages of this consensus approach as compared to, for example, the traditional leader follower approach (see, e.g., [33]) is that a leader follower scheme requires a predefined leader. Since there is no explicit feedback to the leader from the formation in such a scheme and the leader moves independently, the leader may walk away and leave its followers behind. However, the attractiveness of using this leader follower scheme is that a formation coordination problem can simply be reformulated as a typical regulation or tracking control problem. Another approach, which is widely adopted for the formation control, is the artificial potential function approach (e.g., [40]). This approach is based on some Lyapunov-type functions, and these functions represent and realize the inter-sensor interactions and/or the interactions with the environment. An advantage of this approach is that it requires less information to be communicated among sensors and naturally leads to a distributed control law. However, most potential functions-based algorithms suffer from the fact that they do not usually guarantee the convergence to the desired formation pattern.

The goal of this chapter is to design a method for decentralized control of a robotic sensor network that drives the robotic sensors into a flock of the desired geometric shape. To achieve this goal, we use and modify the ideas of Chapter 6. As a first step, we propose a randomized decentralized algorithm to self-deploy all the robotic sensors on vertices of a square grid in an unknown bounded region with unknown obstacles. Furthermore, we also propose a decentralized algorithm that results in forming of a flock of a desired form: After a finite time, all the robotic sensors are deployed inside of a desired geometric shape at vertices of a square grid and different sensors are at different vertices. There are no predefined leaders in the group, the proposed algorithms are fully decentralized, and only local information about the closest neighbors of each sensor is required for the control. Furthermore, we give a mathematically rigorous proof of convergence of the proposed algorithms with probability 1 for any initial positions of the robotic sensors. The performance of the proposed method is also verified by computer simulations.

Some of the results of this chapter are originally published in [100].

The reminder of the chapter is organized as follows. Section 7.2 presents a novel randomized algorithm for self-deployment on a square grid in a bounded region with obstacles. Computer simulations illustrating the proposed algorithm are given in Section 7.3. Our distributed randomized algorithm of flocking with a desired geometric shape is given in Section 7.4. In Section 7.5, we present further illustrative examples. In particular, we consider self-deployment with desired shapes such as interiors of a circle, an ellipse, a rectangle, and a ring.

7.2 Self Deployment on a Square Grid

We consider a network of autonomous mobile robotic sensors in a bounded planar region $\mathscr{R}$ containing a finite number of obstacles $D_1, D_2, \ldots, D_m$; see Fig. 7.1. Our objective is to develop a decentralized algorithm to drive these sensors to vertices of a square grid consisting of equal squares with sides of r while avoiding the obstacles. The value r is known to all sensors a priori.

Let $p_i(\cdot) \in \mathbb{R}^2$ be the vector of the Cartesian coordinates of sensor i. Each mobile sensor has a communication range $r_c \geq \sqrt{2}r$. At a given discrete time $k = 0, 1, 2, \ldots$, each sensor communicates with its surrounding neighbors in a range of $r_c > 0$ for the coordination of their motions. In other words, sensor i has the ability to obtain information of its neighbors in a disk of radius r_c defined by $D_{i, r_c}(k) := \{p \in \mathbb{R}^2 : \|p - p_i(k)\| \leq r_c\}$, where $\|\cdot\|$ denotes the Euclidean norm. On the other hand, the region $\mathscr{R}$ and the obstacles $D_1, \ldots, D_m$ are unknown to the sensors a priori. Each sensor can detect the obstacles and the boundary of the region $\mathscr{R}$ in a range of r_c: for any point p in a range of r_c, the sensor is able to determine whether this point belongs to the region $\mathscr{R}$ and whether this point belongs to the obstacles ($p \in D_s$ for some s). The main assumption on the region $\mathscr{R}$ and the obstacles D_s is as follows.

Assumption 7.1 *The region $\mathscr{R}$ and the obstacles D_s are closed, bounded, and linearly connected sets. For any $s \neq h$, the sets D_s and D_h do not overlap.*

Furthermore, we introduce the set $\mathscr{P}$ consisting of all points p of the region $\mathscr{R}$ that do not belong to D_s for any s.

Definition 7.1 *Consider all possible square grids cutting the plane into equal squares with the sides of r. Let $\mathscr{V}$ be the infinite set of all vertices of this grid. The set $\hat{\mathscr{V}} := \mathscr{V} \cap \mathscr{P}$ is called a square grid set in $\mathscr{P}$; see Fig. 7.1.*

Since each sensor may have restricted communication capabilities, its awareness about the other sensors may be also limited. So the control of the robotic team should be distributed or decentralized in the sense that the movement of each sensor relies on the information of its neighbors, e.g., positions and coordination variables.

Let $\mathscr{N}_i(k)$ be the set of all sensors $j \in \{1, 2, \ldots, n\} \setminus \{i\}$ that at time k belong to the disk $D_{i, r_c}(k)$, and let $|\mathscr{N}_i(k)|$ be the number of elements in $\mathscr{N}_i(k)$. These elements are called *neighbors* of sensor i at time k. For any time $k \geq 0$, the relationships between neighbors are described by a simple undirected graph $G(k)$ with the vertex set $\{1, 2, \ldots, n\}$, where i corresponds to sensor i. The vertices $i \neq j$ of the graph $G(k)$ are connected by an edge if and only if sensors i and j are neighbors at time k. We impose the standard Main Connectivity Assumption (Assumption 6.2 from Chapter 6) on this graph.

Our aim in this section is to develop a decentralized self-deployment algorithm for the robotic sensor network that results in deployment of the sensors at vertices of a square grid set. For any angle θ, we introduce the vectors $n_1(\theta) := (\cos(\theta), \sin(\theta))^T$ and $n_2(\theta) := (-\sin(\theta), \cos(\theta))^T$. These vectors determine the headings of two baselines of a square grid. Any square grid is uniquely defined by an angle θ and a vertex q of the grid. Therefore, any θ and q uniquely define a square set in $\mathscr{P}$, which will be denoted as $\hat{\mathscr{V}}[q, \theta]$. For each robotic sensor i, we introduce the coordination or consensus variables $\theta_i(k) \in \mathbb{R}$ and $q_i(k) \in \mathbb{R}^2$: they represent the current opinion of this sensor on the grid. The sensors start with uncorrelated values of the coordination

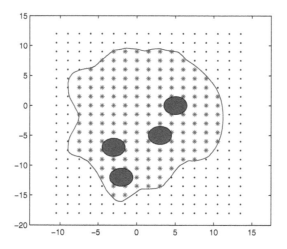

Figure 7.1   Bounded planar region $\mathcal{R}$ containing four obstacles D_1, D_2, D_3, D_4; the set $\mathcal{V}$ is denoted by $\cdot$, and the set $\hat{\mathcal{V}}$ is denoted by $*$.

variables $\theta_i(0)$ and $q_i(0)$, but these variables should eventually converge to common values θ_0 and q_0, which define a common square grid for all sensors.

Assumption 7.2 *The initial values of the coordination variables θ_i satisfy $\theta_i(0) \in [0, \pi)$ for all $i = 1, 2, \dots, n$.*

Let θ be an angle, q and p be points on the plane. The symbol $C[q, \theta](p)$ stands for the closest to p vertex of the square grid set $\hat{\mathcal{V}}[q, \theta]$ (if there are more than one closest vertex, we take any of them). We propose the following rules for updating the consensus variables $\theta_i(k), q_i(k)$ and the coordinates $p_i(k)$ of the robotic sensors:

$$\theta_i(k+1) = \frac{\theta_i(k) + \sum_{j \in \mathcal{N}_i(k)} \theta_j(k)}{1 + |\mathcal{N}_i(k)|},$$

$$q_i(k+1) = \frac{q_i(k) + \sum_{j \in \mathcal{N}_i(k)} q_j(k)}{1 + |\mathcal{N}_i(k)|}, \tag{7.1}$$

$$p_i(k+1) = C[q_i(k), \theta_i(k)]p_i(k). \tag{7.2}$$

The algorithm (7.1), (7.2) can be summarized as follows. The robotic sensors use the variables θ_i to achieve consensus on the square grid heading θ and the variables q_i to achieve consensus on the square grid phase shift. Every sensor is driven to the closest vertex of its current grid set in $\mathcal{P}$.

Remark 7.1 *Initially, the sensors do not have access to a common coordinate system; otherwise, the achievement of consensus would be trivial. So strictly speaking,* the consensus variables $\theta_i(k)$ and $q_i(k)$ of each sensor are given in its own local coordinate frame. However, we assume that each sensor has access to the bearing and distance to each of its neighboring sensors. Using these data, the message from sensor i to a neighboring sensor j is built by sensor i in the form of the consensus variables $\theta_i(k)$ and $q_i(k)$ re-calculated in the right Cartesian coordinate system centered at p_j whose x-axis is aligned with the vector from p_i to p_j. Using this information, each sensor can re-calculate the sums (7.1) in its own coordinate system.

Theorem 7.1 *Suppose that Assumptions 6.2, 7.1, and 7.2 hold and the robotic sensors move according to the decentralized control law (7.1), (7.2). Then there exists a square grid set $\hat{\mathcal{V}}$ with the following property: For any $i = 1, 2, \ldots, n$, there exists a vertex $v \in \hat{\mathcal{V}}$ such that $\lim_{k \to \infty} p_i(k) = v$.*

Proof: Assumption 6.2 and the update law (7.1) guarantee that there exist θ_0 and q_0 such that

$$\theta_i(k) \to \theta_0, \quad q_i(k) \to q_0 \quad \text{as} \quad k \to \infty \qquad \forall i = 1, 2, \ldots, n; \qquad (7.3)$$

see [53]. Furthermore, the update law (7.2) guarantees that $p_i(k+1) \in \hat{\mathcal{V}}[q_i(k), \theta_i(k)]$. This and (7.3) imply that $\lim_{k \to \infty} p_i(k) = v$, where $v \in \hat{\mathcal{V}}[q_0, \theta_0]$. This completes the proof of Theorem 7.1. ∎

The algorithm (7.1), (7.2) drives all the sensors to vertices of some square grid set. However, it does not guarantee that different sensors arrive at distinct vertices of this set. Therefore, we develop the second stage of our algorithm. Now we assume that all the robotic sensors are at the vertices of a square grid set and move from one vertex to another, i.e., $p_i(k) \in \hat{\mathcal{V}}$ for all i, k. Our aim is to drive different sensors to different vertices to achieve the coverage of a larger area and sensors' separation.

For any vertex v of a square grid, the eight closest to v vertices of this square grid are called the *neighbors of v*; see Fig. 7.2. We also need the following.

Assumption 7.3 *The square grid set $\hat{\mathcal{V}}$ is connected: For any $v \in \hat{\mathcal{V}}$, at least one of eight neighbors of v also belongs to $\hat{\mathcal{V}}$.*

Let $\mathcal{S}(p_i(k))$ denote the set consisting of $v = p_i(k)$ and those of eight vertices s of the grid that are neighbors of v, belong to $\hat{\mathcal{V}}$, and $s \neq p_j(k)$ for all $j = 1, 2, \ldots, n$. In other words, $\mathcal{S}(p_i(k))$ is the set consisting of $v = p_i(k)$ and all its "vacant" neighbors in the square grid set. Let $|\mathcal{S}(p_i(k))|$ be the number of elements in $\mathcal{S}(p_i(k))$. It is clear that $1 \leq |\mathcal{S}(p_i(k))| \leq 9$ since $v \in \mathcal{S}(p_i(k))$. Furthermore, let $\mathcal{Q}(p_i(k))$ denote the set consisting of $v = p_i(k)$ and those of eight vertices s of the grid that are neighbors of v and belong to $\hat{\mathcal{V}}$. Let $|\mathcal{Q}(p_i(k))|$ be the number of elements in $\mathcal{Q}(p_i(k))$. It is clear that $2 \leq |\mathcal{Q}(p_i(k))| \leq 9$ because of the inclusion $v \in \hat{\mathcal{Q}}(p_i(k))$ and Assumption 7.3. We also introduce the Boolean variable $b_i(k)$ such that $b_i(k) := 1$ if there exists an index $j \neq i$ for which $p_j(k) = p_i(k)$, and $b_i(k) := 0$ otherwise. In other words, $b_i(k)$ indicates whether sensor i shares a vertex with some other sensor.

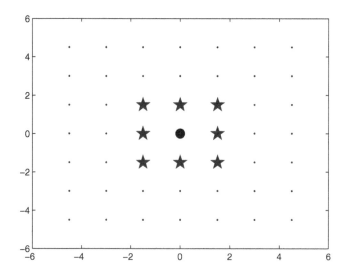

Figure 7.2   A vertex • in a square grid and its eight closest neighbors ⋆.

We propose the following random algorithm:

$$p_i(k+1) = p_i(k) \quad if \quad b_i(k) = 0,$$
$$p_i(k+1) = v \quad with \; probability \quad \frac{1}{|\mathscr{S}(p_i(k))|}$$
$$\forall v \in \mathscr{S}(p_i(k)) \quad if \, (b_i(k) = 1 \quad and \quad |\mathscr{S}(p_i(k))| > 1),$$
$$p_i(k+1) = v \quad with \; probability \quad \frac{1}{|\mathscr{D}(p_i(k))|}$$
$$\forall v \in \mathscr{S}(p_i(k)) \quad if \, (b_i(k) = 1 \quad and \quad |\mathscr{S}(p_i(k))| = 1). \tag{7.4}$$

The control law (7.4) obviously implies that all the sensors do not move whenever they occupy different vertices of the square grid set. This can be achieved only if we have enough vertices in the square grid set.

Assumption 7.4 *Let t_M be the number of vertices in the set $\hat{\mathscr{V}}$. Then $n \le t_M$.*

Theorem 7.2 *Suppose that Assumptions 7.3 and 7.4 hold and the robotic sensors move according to the decentralized control law (7.4). Then with probability 1, there exists a time $k_0 \ge 0$ with the following property: For any $i = 1,2,\dots,n$, there exists a vertex $v_i \in \hat{\mathscr{V}}$ such that $v = p_i(k)$ for all $k \ge k_0$, and $v_i \ne v_j$ for all $i \ne j$.*

Proof: The algorithm (7.4) defines an absorbing Markov chain [44, Ch. 11], which contains a number of absorbing states (i.e., the states that cannot be left with probability 1; in the case at hand, these are the states of the network where different

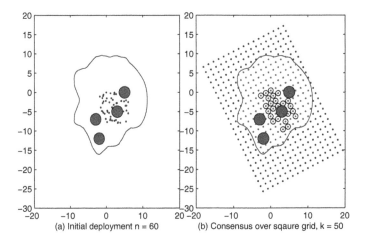

(a) Initial deployment n = 60   (b) Consensus over sqaure grid, k = 50

Figure 7.3 Sensors movement to form a square covering set ($n = 60$): (a) Initial deployment. (b) Consensus over a square grid, $k = 50$.

sensors are located in different vertices of $\hat{\mathcal{V}}$). It is also obvious that these absorbing states can be reached from any initial state with a nonzero probability, and there are no other absorbing states. This implies that with probability 1, one of the absorbing states is reached [44, Ch. 11]. This completes the proof of Theorem 7.2.   ∎

7.3 Illustrative Examples: Square Grid Deployment

To illustrate the proposed algorithm, a series of computer simulations was carried out. The objective was to drive a network of mobile robotic sensors to vertices of a square grid in a bounded region $\mathcal{R}$ with four obstacles, as is shown in Fig. 7.3(a). Sixty sensors, denoted by ·, were initially deployed randomly inside $\mathcal{R}$, as is shown in Fig. 7.3(a). The initial values of the coordination variables $\theta_i(0)$ were randomly distributed over the interval $\left(0, \frac{5}{18}\pi\right)$. The initial values of the consensus variables $q_i(0)$ were randomly distributed near the sensors' initial positions in a square with a side of $5r_s$. The mobile sensors were successfully driven by the control law (7.1), (7.2) to form a square covering set after 50 steps, as is shown in Fig. 7.3(b), where the sensors' positions are denoted by ∘.

Next the algorithm (7.4) was applied to move the sensors to different vertices of the square grid set. Figure 7.4(a) shows the sensors locations after 3 steps and Fig. 7.4(b) shows the sensors' locations after 20 steps when all the sensors are located at different vertices of the square grid set.

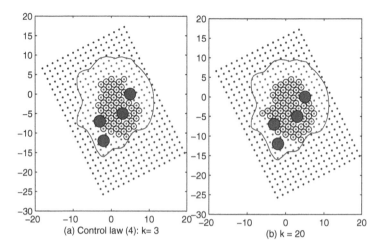

Figure 7.4   Sensor's movement to individual vertices of the square covering set: (a) Sensors moving towards vacant vertices; $k = 3$; (b) $k = 20$.

7.4   Self-Deployment in a Desired Geometric Shape

In this section, we consider a mobile robotic sensor network on a plane without obstacles. Our goal is to develop a decentralized control law that will eventually result in a formation of the sensors with a given desired geometric shape, e.g. circle, ellipse, rectangle, ring, hexagon, etc. At the first stage, we can apply the control law (7.1), (7.2). According to Theorem 7.1, this law drives the sensors to vertices of some square grid, and the sensors achieve a consensus on the square grid direction θ_0 and vertex q_0. The consensus parameters θ_0 and q_0 define a certain orthogonal coordinate system (x, y) on the plane, which is common for all the sensors. In this system, the origin is at q_0 and the x-axis is defined by the direction θ_0. On this coordinate plane, a region $\mathcal{R}$ with a desired geometric shape can be defined by some relationship of the type $f(x, y) \in R$, where $f(\cdot, \cdot)$ is some function and R is some subset of the real axis. We assume that $f(\cdot, \cdot)$ and R are known to each sensor a priori. In particular, we will consider the following geometric shapes and the corresponding relationships:

Circular formation: the relationship $x^2 + y^2 \leq c$;
Ellipsoidal formation: the relationship $ax^2 + by^2 \leq c$;
Rectangular formation: the relationship $\max\{c_x^{-1}|x|; c_y^{-1}|y|\} \leq 1$;
Ring formation: the relationship $c_m \leq x^2 + y^2 \leq c_M$;

Regular hexagon formation: the relationship $\max\limits_{j=0,1,...,5} \left[x\cos\dfrac{\pi j}{3} + y\sin\dfrac{\pi j}{3} \right] \leq c$.

After the first stage of the algorithm from Section 7.2 is completed, all sensors are at vertices of a square grid, but some of them may lie outside of the desired geometric shape. At the second stage, such sensors are moved to the closest square grid vertices that lie inside of the set defined by $f(x,y) \in R$.

After the second stage, all the sensors are at vertices of a square grid set $\hat{\mathscr{V}}$ that are inside the desired geometric shape. At the third stage, we simply apply the control law (7.4) with the region $\mathscr{R}$ defined by the relationship $f(x,y) \in R$. If Assumption 7.4 holds (the number of grid points inside of the desired geometric shape is more or equal to the number of sensors), then Theorem 7.2 implies that after finitely many steps, all the sensors will be deployed inside of the desired geometric shape at different vertices of the square grid.

Thus, we have described the following algorithm composed of three steps:

S1: The control law (7.1), (7.2) is applied.

S2: If some sensors lie outside of the desired geometric shape, they are moved to the closest square grid vertices that are inside of the shape defined by $f(x,y) \in R$.

S3: The control law (7.4) is applied, with the region $\mathscr{R}$ defined by the inclusion $f(x,y) \in R$.

The following theorem is immediate from Theorems 7.1 and 7.2.

Theorem 7.3 *Suppose that all the assumptions of Theorems 7.1 and 7.2 hold, the desired geometric shape is defined by $f(x,y) \in R$, and the robotic network is driven by the three step algorithm S1, S2, S3. Then with probability 1, there exists a time $k_0 \geq 0$ with the following property: For any $i = 1, 2, \ldots, n$, there exists a vertex $v_i \in \hat{\mathscr{V}}$ such that $v = p_i(k)$ for all $k \geq k_0$ and $v_i \neq v_j$ whenever $i \neq j$, where $\hat{\mathscr{V}}$ is the set of all square grid vertices belonging to the shape $f(x,y) \in R$.*

Remark 7.2 *In this chapter, we do not consider the problem of avoiding collisions between mobile sensors. Our algorithm may drive several sensors at the same point p. We assume that in this case, the sensors lie in a neighborhood of p, which is sufficiently large to accommodate several sensors.*

Remark 7.3 *In this chapter, we first apply algorithm (7.1), (7.2) to achieve a consensus, and then we apply the algorithm (7.4) assuming that the sensors now move on a square grid. However, it is possible to combine these two control law in one step, applying at each step both (7.1), (7.2) and (7.4). In this case, to apply (7.4), we need to introduce a parameter $\varepsilon < \frac{3r_s}{2}$ and the neighboring grid vertex is considered to be vacant if there is no a sensor that is closer than ε to it.*

7.5 Illustrative Examples: Various Geometric Shapes

In this section, we present simulation results to illustrate the proposed algorithm. The goal is to move the sensors to form a given geometric formation. We consider 100 mobile robotic sensors initially deployed randomly and denoted by $\cdot$, as is shown in Fig. 7.5(a). The initial values of the coordination variables $\theta_i(0)$ are randomly distributed in a range of $(0, \frac{5}{18}\pi)$. The initial values of the consensus variables $q_i(0)$ are randomly distributed in a square with a side of $5r_s$ close to the sensor's initial positions. The mobile sensors move according to the control law (7.1), (7.2) to reach

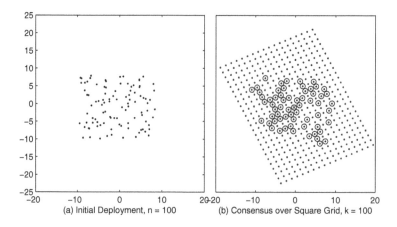

Figure 7.5 Sensors' movement to a square grid: (a) Initial deployment $n = 100$; (b) Consensus over a square grid achieved after 50 steps. The vertex coordinate q_0 is denoted by $\times$.

consensus over the square grid angle θ_0 and its vertex q_0. The sensors eventually converge to the vertices of this square grid set after 50 steps, as is shown in Fig. 7.5(b), with the sensors' positions denoted by '$\circ$' and the "origin" grid vertex q_0 denoted by $\times$. Next we define a region $\mathscr{R}$ with a desired geometric shape on the coordinate plane using the consensus variables θ_0 and q_0.

7.5.1 Circular Formation

First, we consider a circular region $\mathscr{R}$ shown in Fig. 7.6(a), with a radius of $6r_s$, which is known to all sensors. The sensors that are initially outside of $\mathscr{R}$ first move inside this region, as is shown in Fig. 7.6(b). After this, the sensors use the control law given in (7.4) to move to individual vertices of the grid set inside $\mathscr{R}$ to obtain the desired circular formation. Figure 7.6(c) shows the sensors' locations after 3 steps and Fig. 7.6(d) shows the sensors' locations after 20 steps when all the sensors are located at different vertices of the square grid inside the circular region $\mathscr{R}$.

7.5.2 Ellipse Formation

Second, we define an ellipsoidal region with a radius of $4.6r_s$ along the x-axis and a radius of $7.3r_s$ along the y-axis, as is shown in Fig. 7.7 (a). The sensors that are initially outside this region first move inside it, as is shown in Fig. 7.7 (b). Then using the control law (7.4), the sensors move to individual vertices of the square grid inside the region at hand to reach the desired ellipsoidal formation. Figure 7.7(c) shows the sensors' locations after 3 steps and Fig. 7.7(d) shows the sensors' locations after 20

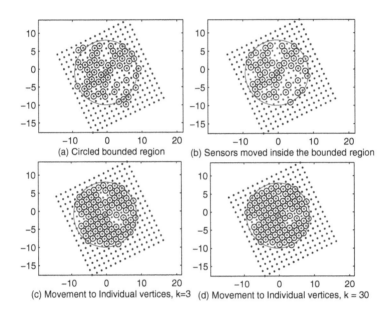

Figure 7.6 Sensors' movement to obtain a circular formation: (a) Circular region and sensors positions over a square grid set; (b) Sensors move inside $\mathcal{R}$; (c) Sensors' movement to individual vertices of the square grid set inside the region $\mathcal{R}$, $k = 3$; (d) $k = 20$.

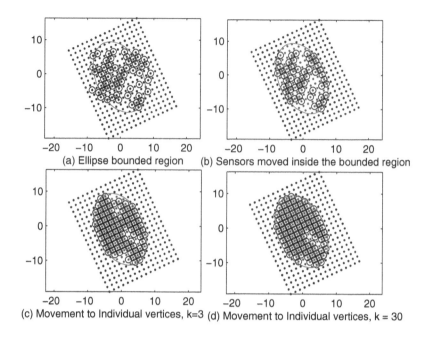

Figure 7.7   Sensors' movement to obtain an ellipse formation: (a) Ellipsoidal region $\mathscr{R}$ and sensors positions over a square grid set; (b) Sensors move inside $\mathscr{R}$; (c) Sensors' movement to individual vertices of the square grid set inside the region $\mathscr{R}$, $k = 3$; (d) $k = 20$.

steps when all the sensors are located at different vertices of the square grid inside the ellipsoidal region.

7.5.3   Rectangular Formation

Third, a rectangular region is defined with sides of $8r_s$ and $13.3r_s$, as is shown in Fig. 7.8 (a). The sensors that are initially outside the rectangular region move inside it, as is shown in Fig. 7.8 (b). Then using the control law (7.4), the sensors move to individual vertices of the square grid inside the region at hand to reach the desired rectangular formation. Figure 7.8(c) shows the sensors' locations after 3 steps and Fig. 7.8(d) shows the sensors' locations after 20 steps when all the sensors are located at different vertices of the square grid inside the rectangular region.

7.5.4   Ring Formation

A ring is defined with an outer radius of $6.3r_s$ and an inner radius of $4r_s$, as is shown in Fig. 7.9 (a). The sensors that are initially outside the ring then move inside it, as is

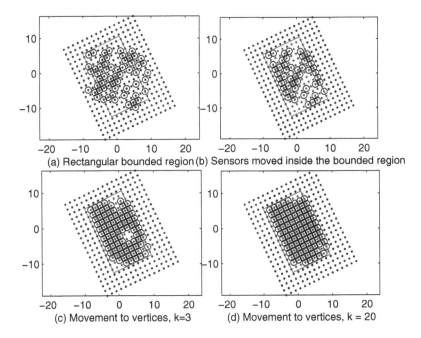

(a) Rectangular bounded region (b) Sensors moved inside the bounded region

(c) Movement to vertices, k=3   (d) Movement to vertices, k = 20

Figure 7.8   Sensors' movement to obtain a rectangular formation: (a) Rectangular region $\mathcal{R}$ and sensors positions over a square grid set; (b) Sensors move inside $\mathcal{R}$; (c) Sensors' movement to individual vertices of the square grid set inside the region $\mathcal{R}$, $k = 3$; (d) $k = 20$.

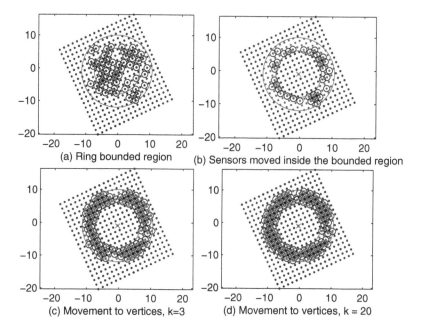

Figure 7.9   Sensors' movement to obtain a ring formation: (a) Ring region $\mathscr{R}$ and sensors positions over a square grid set; (b) Sensors move inside $\mathscr{R}$; (c) Sensors' movement to individual vertices of the square grid set inside the region $\mathscr{R}$, $k = 3$; (d) $k = 20$.

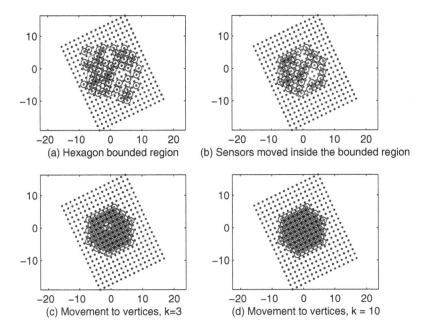

Figure 7.10   Sensors' movement to obtain a regular hexagon formation: (a) Hexagon region $\mathscr{R}$ and sensors positions over a square grid set; (b) Sensors move inside $\mathscr{R}$; (c) Sensors' movement to individual vertices of the square grid set inside the region $\mathscr{R}$, $k = 3$; (d) $k = 10$.

shown in Fig. 7.9 (b). Then using the control law (7.4), the sensors move to individual vertices of the square grid inside the ring $\mathscr{R}$ to reach the desired ring formation. Figure 7.9(c) shows the sensors' locations after 3 steps and Fig. 7.9(d) shows the sensors' locations after 20 steps when all the sensors are located at different vertices of the square grid inside the ring.

7.5.5 Regular Hexagon Formation

Lastly, a regular hexagon is defined with a side of $4.57r_s$, as is shown in Fig. 7.10 (a). The sensors that are initially outside the hexagon then move inside it, as is shown in Fig. 7.10(b). Then using the control law (7.4), the sensors move to individual vertices of the square grid inside the hexagon $\mathscr{R}$ to reach the desired hexagon formation. Figure 7.10(c) shows the sensors' locations after 3 steps and Fig. 7.10(d) shows the sensors' locations after 10 steps when all the sensors are located at different vertices of the square grid inside the hexagon.

CHAPTER 8

MOBILE SENSOR AND ACTUATOR NETWORKS: ENCIRCLING, TERMINATION AND HANNIBAL'S BATTLE OF CANNAE MANEUVER

8.1 Introduction

This chapter introduces a new type of coverage problems which is referred to as encircling coverage. In the encircling coverage problem studied here, the goal is to deploy a network of mobile robotic sensors around a planar region $\mathscr{R}$. Precisely, we wish to deploy the sensors in the region $\mathscr{R}(d_{min}, d_{max})$ consisting of all points at distances between d_{min} and d_{max} to the region $\mathscr{R}$. Moreover, every point $\mathscr{R}(d_{min}, d_{max})$ should be sensed by at least one of the mobile sensors. The region $\mathscr{R}$ is bounded, connected, and of an arbitrary shape, which is not known to the sensors a priori. In practice, the region $\mathscr{R}$ is often related to an unknown scalar environmental field. This field may represent the strength of a spatially distributed signal, e.g., an electromagnetic or acoustic one, the concentration of a chemical, physical, or biological agent, the distribution of a physical quantity, such as thermal, magnetic, electric, or optical field distributions, etc. Examples of missions where such an encircling coverage is of interest include environmental studies, detecting and localizing the sources of hazardous chemicals leakage or vapour emission, sources of pollutants and plumes, hydrothermal vents, environmental monitoring of disposal sites on the deep ocean floor [54], and sea floor surveying for hydrocarbon exploration [9]. Furthermore, the

Decentralized Coverage Control Problems for Mobile Robotic Sensor and Actuator Networks. **113**
By Andrey V. Savkin, Teddy M. Cheng, Zhiyu Xi, Faizan Javed, Alexey S. Matveev, and Hung Nguyen. Copyright © 2015 by the Institute of Electrical and Electronics Engineers, Inc.

region $\mathcal{R}$ may be moving and deformable; in this case, the set $\mathcal{R}(d_{min}, d_{max})$ also moves and changes its shape, therefore, the sensors should move to follow $\mathcal{R}$.

We propose a decentralized randomized control algorithm that drives the network of mobile robotic sensors to form a grid covering $\mathcal{R}(d_{min}, d_{max})$ and consisting of equal equilateral triangles. A major benefit of deploying sensors in the triangular lattice pattern is that it is asymptotically optimal in terms of minimum number of sensors required for the complete coverage of an arbitrary bounded planar set, as it was demonstrated in Chapter 6 based on the famous mathematical result of Kershner [56]. The computationally efficient approach of this chapter is based on the result of Chapter 6 and leads to a solution that is very close to the optimum provided that the number n of mobile robotic sensors is large.

We also introduce the problem of termination of a moving environmental region by a mobile robotic sensor and actuator network. In real-world applications, this moving region may represent an oil spill or an area contaminated with a hazardous chemical or biological agent. In this problem, we assume that the nodes in the network are mobile autonomous robots or vehicles. Furthermore, they are equipped with not only sensors and also actuators that release a neutralizing chemical to control the shape of the polluted region. In other words, the nodes of the networks are not just capable of measuring the moving region in their neighborhoods, but also capable of terminating parts of this region. Also, we consider the case when the field can terminate the moving sensors/actuators as well. In such control applications, actuation plays a major role. We apply our encircling coverage algorithm to this problem. We show that in many realistic situations, our algorithm achieves complete termination of the moving hazardous field.

Our algorithm for encircling/termination of a moving region is partially inspired by the famous Hannibal double-envelopment maneuver during the Battle of Cannae in which the army of Carthage under Hannibal decisively defeated a numerically superior army of the Roman Republic in 216 BC; see, e.g., [43]. Our computer simulations show that the proposed algorithm encircles and terminates moving regions in a patter similar to encirclement and annihilation of the Roman army by the Carthaginians.

Unlike many coverage algorithms proposed in this area, the randomized algorithm presented in this chapter is theoretically verified. In particular, we give a mathematically rigorous proof of convergence of our algorithm with probability 1 for any initial positions of the mobile sensors. The performance of the proposed method is also verified by computer simulations.

Some of the results of this chapter are originally published in [110].

The remainder of the chapter is organized as follows. Section 8.2 presents the statement of the problem of encircling coverage. Our randomized algorithm of network self-deployment and its mathematical analysis are given in Section 8.3. Section 8.4 describes the problem of region termination by a mobile robotic sensor and actuator network and adapts the proposed encircling algorithm to this problem. Finally, Section 8.5 presents computer simulations of the proposed algorithms.

8.2 Encircling Coverage of a Moving Region

Our first objective is to design a decentralized algorithm to drive a network of autonomous mobile sensors to encircle a moving and possibly deforming bounded two-dimensional region $\mathcal{R}(k) \subset \mathbb{R}^2$. The network consists of n mobile sensors, labelled 1 through n. Let $p_i(\cdot) \in \mathbb{R}^2$ be the vector of the Cartesian coordinates of sensor i. Each mobile sensor has a sensing range of $r_s > 0$ and a communication range of $r_c \geq \sqrt{3}r_s$: At any time $k > 0$, each sensor communicates with its surrounding neighbors in a range of $r_c > 0$ for the coordination of their motions. In other words, sensor i has the ability to obtain information on its neighbors in a disk of radius r_c defined by $D_{i,r_c}(k) := \{p \in \mathbb{R}^2 : \|p - p_i(k)\| \leq r_c\}$, where $\|\cdot\|$ denotes the Euclidean norm. Meanwhile, the region $\mathcal{R}(k)$ is unknown to the sensors a priori. In terms of detection, each sensor can detect or sense any object in a disk of radius r_s.

To proceed, we define the distance **dist**$(p,\mathcal{R})$ between the point p and the set $\mathcal{R}$

$$\mathbf{dist}(p,\mathcal{R}) := \inf_{x \in \mathcal{R}} \|p - x\|$$

and the distance **dist**$(\mathcal{R}_1,\mathcal{R}_2)$ between the sets $\mathcal{R}_1$ and $\mathcal{R}_2$

$$\mathbf{dist}(\mathcal{R}_1,\mathcal{R}_2) := \inf_{x_1 \in \mathcal{R}_1, x_2 \in \mathcal{R}_2} \|x_1 - x_2\|.$$

Let

$$d_{max} > d_{min} > 2r_s \tag{8.1}$$

be some given constants. We introduce the set $\mathcal{R}(k,d_{min},d_{max})$ consisting of all points p such that

$$d_{min} \leq \mathbf{dist}(p,\mathcal{R}(k)) \leq d_{max}. \tag{8.2}$$

Also, we consider the r_s-neighborhood $\mathcal{R}(k,d_{min},d_{max},r_s)$ of $\mathcal{R}(k,d_{min},d_{max})$. As in Chapter 6, we impose only a minor assumption on the region.

Assumption 8.1 *For any $k = 0,1,2,\ldots$, the region $\mathcal{R}(k,d_{min},d_{max})$ is bounded and connected.*

We assume that each sensor has the capability to measure the distance **dist**$(p_i(k), \mathcal{R}(k))$ between its current position and the set $\mathcal{R}(k)$ at any time k. Moreover, we assume that the constants d_{min} and d_{max} are known to the sensors a priori.

Definition 8.1 *A finite set of points $\mathcal{W}$ is said to be a* complete blanket coverage *of $\mathcal{R}(k,d_{min},d_{max})$ if for any $p \in \mathcal{R}(k,d_{min},d_{max})$, there exists a point $w \in \mathcal{W}$ such that $\|p - w\| \leq r_s$.*

We stress that some points w of $\mathcal{W}$ may be outside of $\mathcal{R}(k,d_{min},d_{max})$.

Definition 8.2 *Consider all possible triangular grids cutting the plane into equilateral triangles with the sides of $\sqrt{3}r_s$. Let $\mathcal{V}$ be one of these triangular grids. The set $\hat{\mathcal{V}}(k) := \mathcal{V} \cap \mathcal{R}(k,d_{min},d_{max},r_s)$ is called a* triangular covering set *of $\mathcal{R}(k,d_{min},d_{max})$; see Fig. 8.1.*

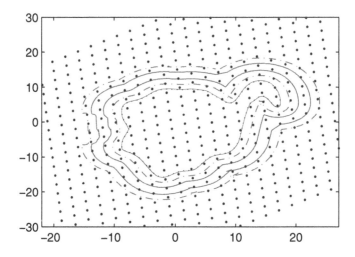

Figure 8.1   Bounded two-dimensional region $\mathscr{R}$; the set $\mathscr{R}(k,d_{min},d_{max})$ is inside the solid lines and $\mathscr{R}(k,d_{min},d_{max},r_s)$ is inside the dashed lines.

Any triangular covering set of $\mathscr{R}(k,d_{min},d_{max})$ is a complete blanket coverage of $\mathscr{R}(k,d_{min},d_{max})$. Indeed, for any $p \in \mathscr{R}(k,d_{min},d_{max})$, there obviously exists a vertex $v \in \mathscr{V}$ such that $\|p - v\| \leq r_s$ (v is the closest to p vertex of the triangular grid). Hence, $v \in \mathscr{R}(k,d_{min},d_{max},r_s)$. This implies that $v \in \hat{\mathscr{V}}(k)$.

So if the sensor network is deployed so that there is a sensor at each point v of some triangular covering set, a complete detection coverage of $\mathscr{R}(k,d_{min},d_{max})$ is achieved.

The following theorem obviously follows from Theorem 6.1 in Chapter 6 and shows that triangular coverages are asymptotically optimal: For small r_s, they contain almost the minimum possible number of vertices.

Theorem 8.1 *Suppose that Assumption 8.1 holds. For any $r_s > 0$, let $\hat{\mathscr{V}}(r_s)$ be a triangular covering set of $\mathscr{R}(k,d_{min},d_{max})$, and let $t(r_s)$ be the number of points in $\hat{\mathscr{V}}(r_s)$. Also, let $m(r_s)$ be the minimum possible number of points for all complete blanket coverages of $\mathscr{R}(k,d_{min},d_{max})$. Then $\lim_{r_s \to 0} \frac{t(r_s)}{m(r_s)} = 1$.*

Since each mobile sensor may have a restricted communication and detection capabilities to reduce the cost of operation, it also may have a limited information about other sensors in the group. This information is local and limited to its neighbors, as each sensor can only communicate with or observe its neighbors, not all the sensors in the network. In practice, the locations of neighboring sensors can be estimated from the range-only measurements using robust Kalman estimation methods; see, e.g., [88, 102, 103].

As a result, the control law of each mobile sensor should be distributed or decentralized in the sense that the movement of each sensor relies on the information of its neighbors, e.g., positions and coordination variables.

Let $\mathcal{N}_i(k)$ be the set of all sensors $j \in \{1, 2, \ldots, n\} \setminus \{i\}$ that at time k belong to the disk $D_{i,r_c}(k)$, and let $|\mathcal{N}_i(k)|$ be the number of elements in $\mathcal{N}_i(k)$. These elements are called the *neighbors* of sensor i at time k. For any $k \geq 0$, the relationships between neighbors are described by a simple undirected graph $G(k)$ with the vertex set $\{1, 2, \ldots, n\}$, where i corresponds to sensor i. The vertices $i \neq j$ of the graph $G(k)$ are connected by an edge if and only if the sensors i and j are neighbors at time k. In this chapter, we again impose the standard Main Connectivity Assumption (Assumption 6.2 from Chapter 6) on the graph sequence $G(k)$.

We recall that the region $\mathcal{R}(k)$ may move or deform. We assume that the rate of motion and deformation is not too large by imposing the following.

Assumption 8.2 *For any $k = 0, 1, 2, \ldots$, the following inequality holds:*

$$\mathbf{dist}(\mathcal{R}(k), \mathcal{R}(k+1)) \leq r_s.$$

Our aim is to develop a distributed algorithm for the mobile sensor network whose application results in encircling the moving region $\mathcal{R}(k)$ by self-deployment of the sensors at the vertices of a triangular covering set of $\mathcal{R}(k, d_{min}, d_{max})$.

8.3 Randomized Encircling Algorithm

For any angle θ, we introduce the vectors $n_1(\theta) := (\cos(\theta), \sin(\theta))^T$ and $n_2(\theta) := (\cos(\theta + \frac{\pi}{3}), \sin(\theta + \frac{\pi}{3}))^T$. These vectors, together with the vector $n_1(\theta) - n_2(\theta)$, determine the headings of three baselines of a triangular grid. Any triangular grid is uniquely defined by an angle θ and a point q, which is a vertex of the grid. Therefore, any θ and q uniquely define a triangular covering set of $\mathcal{R}(k, d_{min}, d_{max})$, which will be denoted as $\mathcal{V}[q, \theta](k)$. For each mobile sensor i, we introduce the coordination or consensus variables $\theta_i(k)$ and $q_i(k)$, which represent the current opinion of this sensor on the grid. The sensors may start with different opinions given by $\theta_i(0)$ and $q_i(0)$; however, these variables should eventually converge to some consensus values θ_0 and q_0, which define a common triangular grid for all sensors.

Assumption 8.3 *The initial values of the coordination variables θ_i satisfy $\theta_i(0) \in [0, \pi)$ for all $i = 1, 2, \ldots, n$.*

To proceed, we need some notation. Let θ be an angle, and let q, p be points on the plane. Then $C[q, \theta, k](p)$ denotes the closest to p vertex of the triangular covering set $\mathcal{V}[q, \theta](k)$ (if there is more than one closest vertex in $\mathcal{V}[q, \theta](k)$, we take any of them). We propose the following rules for updating the consensus variables

$\theta_i(k), q_i(k)$ and the sensors' coordinates $p_i(k)$:

$$\theta_i(k+1) = \frac{\theta_i(k) + \sum_{j \in \mathcal{N}_i(k)} \theta_j(k)}{1 + |\mathcal{N}_i(k)|},$$

$$q_i(k+1) = \frac{q_i(k) + \sum_{j \in \mathcal{N}_i(k)} q_j(k)}{1 + |\mathcal{N}_i(k)|}; \qquad (8.3)$$

$$p_i(k+1) = C[q_i(k), \theta_i(k), k](p_i(k)). \qquad (8.4)$$

The algorithm (8.3), (8.4) can be summarized as follows. The mobile sensors use the variables θ_i to achieve the consensus on the triangular grid heading θ and use the variables q_i to achieve the consensus on the triangular grid phase shift. Each sensor is moved to the closest vertex of its current triangular covering set.

Remark 8.1 *The sensors initially do not have a common coordinate system; otherwise, achievement of consensus would be trivial. So the consensus variables $\theta_i(k)$ and $q_i(k)$ of every sensor are in fact with respect to its own local coordinate frame. However, we assume that each sensor has access to the bearing of and distance to each of its neighboring sensors. On the basis of these data, the message from sensor i to a neighboring sensor j formed by sensor i at time k consists of the consensus variables $\theta_i(k)$ and $q_i(k)$ re-calculated in the coordinate system centered at p_j whose x-axis is aligned with the vector from p_i to p_j. Using this information, each sensor can re-calculate the sums (8.3) in its own coordinate system.*

Theorem 8.2 *Suppose that Assumptions 6.2, 8.1, 8.3 hold and the mobile sensors move according to the decentralized control law (8.3), (8.4). Then there exists a triangular grid $\mathcal{V}$ such that the sequence $\widehat{\mathcal{V}}(k)$ (introduced in Definition 8.2) has the following property: For any $i = 1, 2, \dots, n$, there exists a sequence $v_i(k) \in \widehat{\mathcal{V}}(k)$ such that $\lim_{k \to \infty}[p_i(k) - v_i(k)] = 0$.*

Proof: Assumption 6.2 and the update law (8.3) guarantee that there exist θ_0 and q_0 such that

$$\theta_i(k) \to \theta_0, \quad q_i(k) \to q_0 \quad \text{as} \quad k \to \infty \qquad \forall i = 1, 2, \dots, n; \qquad (8.5)$$

see [53]. Furthermore, the update law (8.4) guarantees that $p_i(k+1) \in \widehat{\mathcal{V}}[q_i(k), \theta(k)](k)$. This and (8.5) imply that $\lim_{k \to \infty}[p_i(k) - v_i(k)] = 0$ for some sequence $v_i(k) \in \widehat{\mathcal{V}}(k)$, which completes the proof of Theorem 8.2. ■

The algorithm (8.3), (8.4) drives all the sensors to vertices of triangular covering sets. However, it does not guarantee that the sensors will occupy all the vertices. Therefore, we develop the second stage of our algorithm. Now we assume that all the mobile sensors are at the vertices of a triangular grid set and move from one vertex to another, i.e., $p_i(k) \in \widehat{\mathcal{V}}(k)$ for all i, k. Our aim is to occupy all the vertices of the covering sets $\widehat{\mathcal{V}}(k)$ to achieve the complete coverage. Let $\mathscr{S}(p_i(k), k)$

denote the set consisting of $v = p_i(k)$, and the vertices s of the grid that satisfy the following properties: They are neighbors of v, belong to $\hat{\mathcal{V}}(k)$, and $s \neq p_j(k)$ for all $j = 1, 2, \ldots, n$. In other words, $\mathcal{S}(p_i(k), k)$ is the set consisting of $v = p_i(k)$ and all its "vacant" neighbors in the triangular covering set. Let $|\mathcal{S}(p_i(k), k)|$ be the number of elements in $\mathcal{S}(p_i(k))$. It is clear that $1 \leq |\mathcal{S}(p_i(k), k)| \leq 7$ since $v \in \hat{\mathcal{V}}$.

We propose the following random algorithm:

$$p_i(k+1) = s \quad with \;\; probability \;\; \frac{1}{|\mathcal{S}(p_i(k), k)|} \qquad \forall s \in \mathcal{S}(p_i(k), k). \qquad (8.6)$$

Now we are going to give a mathematical analysis of this algorithm for the case of a steady region

$$\mathcal{R}(k) \equiv \mathcal{R} \qquad \forall k = 0, 1, 2, \ldots . \qquad (8.7)$$

In this case, $\hat{\mathcal{V}}(k) \equiv \hat{\mathcal{V}}$ for all k. Furthermore, the control law (8.6) obviously implies that all the sensors stop when all the vertices of the triangular covering set are occupied. This can be achieved only if we have enough mobile sensors.

Assumption 8.4 *Let t_M be the maximum possible number of vertices in the set $\hat{\mathcal{V}}$. Then $n \geq t_M$.*

Theorem 8.3 *Consider a steady region (8.7) and suppose that Assumption 8.4 holds and the mobile sensors are driven by the decentralized control law (8.6). Then with probability 1, there exists a time $k_0 \geq 0$ such that for any $v \in \hat{\mathcal{V}}$, the relationship $v = p_i(k)$ holds for some $i = 1, 2, \ldots, n$ and all $k \geq k_0$.*

Proof: The algorithm (8.6) defines an absorbing Markov chain [44, Ch. 11], which contains absorbing states (i.e., states that cannot be left with probability 1); in the case at hand, these are the states of the network, where all the vertices of $\hat{\mathcal{V}}$ are occupied. These absorbing states can obviously be reached from any initial state with a nonzero probability, and there are no other absorbing states. This implies that with probability 1, one of the absorbing states is reached [44, Ch. 11]. This completes the proof of Theorem 8.3. ∎

Remark 8.2 *In this chapter, we first apply algorithm (8.3), (8.4) to achieve a consensus, and then we apply the algorithm (8.6) assuming that the sensors now move on a triangular grid. However, it is possible to simultaneously execute these two algorithms by applying (8.3), (8.4), and (8.6) at each step. In this case, we need to introduce a parameter $\varepsilon < \frac{3r_s}{2}$ to apply (8.6), and a neighboring grid vertex is considered to be vacant if there is no sensor that is closer than ε to it.*

8.4 Termination of a Moving Region by a Sensor and Actuator Network

In this section, we model a moving environmental region $\mathcal{R}(k)$ by a discrete set of points $r_j(k)$ on the plane, where $j = 1, 2, \ldots, m$. In other words, we consider a moving set that consists of m mobile points, labelled 1 through m. Here $r_j(\cdot) \in \mathbb{R}^2$ is the vector of the Cartesian coordinates of point j.

We assume that all the points $r_j(k)$ are at vertices of a square grid with a side $\varepsilon_0 \in (0, r_s)$ at time $k = 0$, and that these points move according to the following law:

$$r_j(k+1) = r_j(k) + v_0 + v_j(k) \quad \forall j = 1, \ldots, m, \tag{8.8}$$

where v_0 is a given constant vector and $v_j(k)$ is the noise.

In this section, we consider a network of mobile nodes or agents that are endowed with not only sensing but also actuating capacity. This means that the nodes are not just capable of measuring the moving region in their neighborhoods but also capable of terminating the moving points representing this region. In other words, the mobile nodes carry not just sensors but actuators releasing, e.g., a neutralizing chemical so that the shape of the polluted region is controlled. In the real world, the moving region may represent an oil spill or hazardous chemical or biological field. The sensor/actuator nodes terminate this field. Also, we consider the case when the field can terminate the moving nodes as well.

Specifically, let $R_T > 0$ be a given constant, which is called the *termination radius*. Also, let $b_p \geq b_r > 0$ be given constants, and let $N_p(x,k)$ denote the number of the mobile sensors $p_i(k)$ inside the closed disk of radius R_T centred at the point x. Similarly, let $N_r(x,k)$ denote the number of the mobile points $r_j(k)$ inside the closed disk of radius R_T centred at the point x.

At any time $k > 0$, some of the mobile sensors i and some of the mobile points j are terminated according to the following rules:

Sensor termination rule: The sensor $p_i(k)$ is terminated if

$$\frac{N_r(p_i(k),k)}{N_p(p_i(k),k)} > b_p. \tag{8.9}$$

Point termination rule: The point $r_j(k)$ is terminated if

$$\frac{N_r(r_j(k),k)}{N_p(r_j(k),k)} < b_r. \tag{8.10}$$

Now the set $\mathscr{R}(k)$ consists of all the points $r_j(k)$ that have not been terminated by the time k. We examine the encircling algorithm (8.3), (8.4), (8.6) acting in parallel with the termination rules (8.9), (8.10).

8.5   Illustrative Examples

In this section, we present computer simulation results to illustrate the proposed algorithms. Our objective is to obtain an encircling coverage to completely cover the region $\mathscr{R}(k, d_{min}, d_{max})$ shown in Fig. 8.1. We consider 100 mobile sensors initially deployed randomly at the lower right side of $\mathscr{R}(k, d_{min}, d_{max})$, as is shown in Fig. 8.2. The dashed lines in Fig. 8.2 represent the boundary of $\mathscr{R}(k, d_{min}, d_{max}, r_s)$. The initial values of the coordination variables $\theta_i(0)$ are randomly distributed in the range of $\left(0, \frac{5}{18}\pi\right)$. The initial values of the consensus variables $q_i(k)$ are randomly distributed in a square with a side of $8.5r_s$. In the first 40 steps of the current simulation,

$\mathcal{R}(k)$ moves towards the top left corner, with its horizontal and vertical coordinates shrinking and stretching, respectively. From the 41st step, $\mathcal{R}(k)$ moves towards the top right corner, with its horizontal and vertical coordinates stretching and shrinking, respectively. The speed of motion and the rate of deformation are relatively small, as was suggested in Assumption 8.2. In the first 30 steps, the sensors move according to the algorithm (8.3), (8.4) to form a triangular covering set. Figures 8.3 and 8.4 show sensors' movement to form a complete blanket coverage, where the sensors' positions are denoted by ○. Consensus over a triangular grid after 20 steps is shown in Fig. 8.3. From the 31st step, the algorithm (8.6) was applied to obtain a complete blanket coverage. Within each step, (8.3) and (8.4) are applied prior to (8.6). Figure 8.4 shows the sensors' locations after 20 steps, and Fig. 8.5 shows the sensors' locations after another 50 steps when all the vertices of the triangular covering set have been occupied and the region $\mathcal{R}(k, d_{min}, d_{max})$ has been completely covered by the sensors.

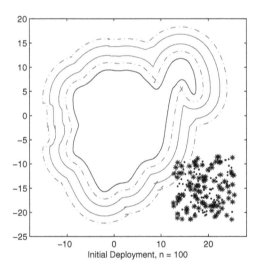

Figure 8.2   Initial deployment.

Figures 8.6—8.11 display simulation results of the termination law proposed in Section 8.4. Figure 8.6 shows the initial deployment. The region $\mathcal{R}$ consists of 200 mobile points represented by dots. A hundred of mobile sensors are denoted by black points, R_T is now chosen as $2r_s, \varepsilon_0 = 0.5r_s, b_p = 8$, and $b_r = 7$.

In the first ten steps of simulation, the algorithm (8.3), (8.4) is applied and from $k = 10$, the algorithm (8.6) is applied to obtain a complete blanket coverage. Within each step, (8.3), (8.4) are applied prior to (8.6). The left panel of Fig. 8.7 shows the sensors' locations after 10 steps, and the right panel shows the sensors' locations after another 10 steps. The vertices of the triangular grid that fall into $\mathcal{R}(k, d_{min}, d_{max}, r_s)$ are denoted by the dots. Fig. 8.8 shows that the region $\mathcal{R}$ has been encircled by mo-

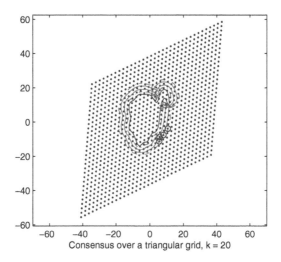

Figure 8.3   Consensus over a triangular grid, $k = 20$.

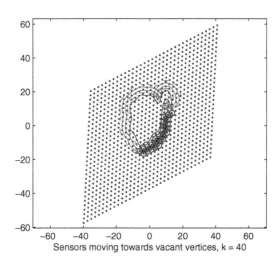

Figure 8.4   Sensors moving towards vacant vertices, $k = 40$.

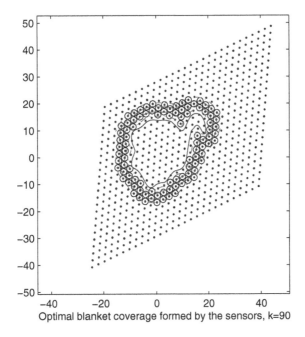

Optimal blanket coverage formed by the sensors, k=90

Figure 8.5   Blanket coverage, $k = 90$.

Table 8.1   Effect of the choice of b_p and b_r

b_p	b_r	Termination	No. of steps needed
8	7	Region $\mathscr{R}$	44
6	5	Region $\mathscr{R}$	55
4	3	Region $\mathscr{R}$	100
2	2	Sensors	95
2	1	Sensors	52

bile sensors, and some of mobile points forming the region $\mathscr{R}$ have been terminated according to (8.6) at $k = 32$. Eventually, the region $\mathscr{R}$ vanishes and all mobile points are terminated at the step $k = 44$, as is shown in Fig. 8.9.

Figures 8.10 and 8.11 display the simulation results with the same initial deployment in the case where b_p and b_r in the termination law are set to be $b_p = 2$ and $b_r = 1$. It is observed that after 52 steps, all sensors are terminated. Table 8.1 provides a better insight into the effect from the choice of b_p and b_r.

Figure 8.12 shows the numbers of steps performed until termination of the region in the case where b_p is set as infinity (so the sensors are never terminated) with

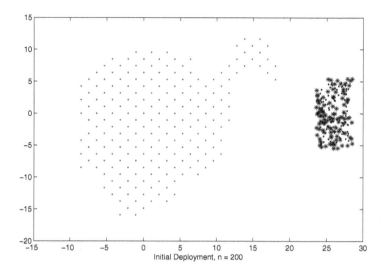

Figure 8.6   Initial deployment, $n = 200$. The region $\mathscr{R}$ consists of the mobile points represented by the red dots.

Table 8.2   Number of steps decreases as b_r increases

b_p	b_r	Termination	No. of steps needed
∞	9	Region $\mathscr{R}$	32
∞	8	Region $\mathscr{R}$	37
∞	7	Region $\mathscr{R}$	43
∞	6	Region $\mathscr{R}$	44
∞	5	Region $\mathscr{R}$	55
∞	4	Region $\mathscr{R}$	72
∞	3	Region $\mathscr{R}$	82

different values of b_r. As is expected, the number of steps decreases as b_r increases. These results are specified in Table 8.2.

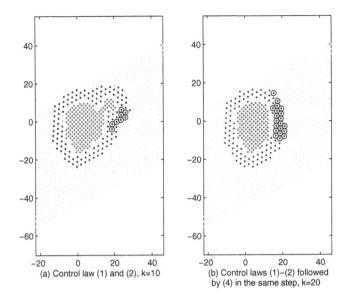

(a) Control law (1) and (2), k=10

(b) Control laws (1)–(2) followed
by (4) in the same step, k=20

Figure 8.7　The algorithm (8.3), (8.4) and (8.6) with $b_r = 7, b_p = 8$.

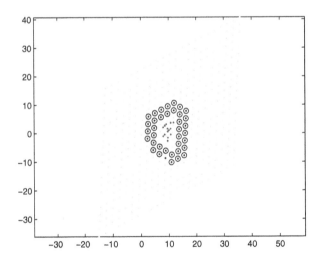

Figure 8.8　The region $\mathscr{R}$ encircled by mobile sensors, $b_r = 7, b_p = 8$.

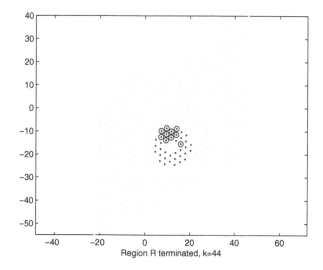

Figure 8.9   The region $\mathcal{R}$ is terminated, $k = 44$.

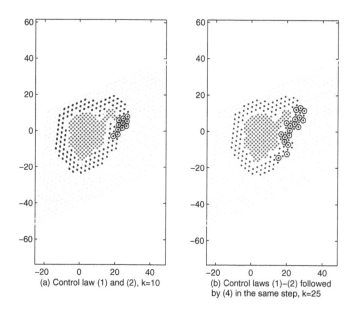

Figure 8.10   The algorithm (8.3), (8.4) and (8.6) with $b_r = 1, b_p = 2$.

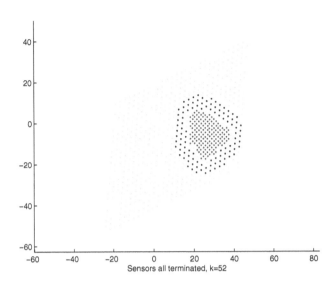

Figure 8.11   All sensors are terminated, $k = 52$.

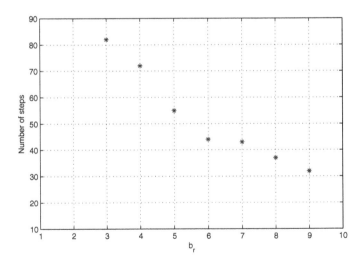

Figure 8.12   The number of steps required to terminate the region $\mathscr{R}$ with $b_p = \infty$ and various values of b_r.

CHAPTER 9

ASYMPTOTICALLY OPTIMAL BLANKET COVERAGE BETWEEN TWO BOUNDARIES

9.1 Introduction

In this chapter, we address another version of the blanket coverage problem using a mobile robotic sensor network. The blanket coverage problem of Chapter 6 was to deploy a network of mobile sensors so that a given two-dimensional region is completely covered and every point in the region is sensed by at least one sensor. In the blanket coverage problem of this chapter, we do not have a fixed two-dimensional region. Our objective is to find and cover a bounded region between two given lines $\mathscr{W}_1$ and $\mathscr{W}_2$; these may be, e.g., the walls or fences of a corridor. Given two points, say L_1 and L_2, on the line $\mathscr{W}_1$, our goal is to deploy a group of mobile autonomous sensors to form a sensor lattice that fully covers the region $\mathscr{R}$ defined by L_1 and L_2 (see Fig. 9.1), whereas the initial deployment may be arbitrary. In Fig. 9.1(a), the mobile sensors are randomly dispatched near L_1 at the initial deployment, which does not provide blanket coverage of $\mathscr{R}$. To address the coverage problem, we propose a set of control laws that drives the network of sensors to form a sensor lattice. In particular, we are interested in deploying the sensors in the so-called triangular lattice pattern, which is composed of equilateral triangles; see Fig. 9.1(b). As is illustrated in Fig. 9.1(b), every point in $\mathscr{R}$ is covered by at least one sensor, and

Decentralized Coverage Control Problems for Mobile Robotic Sensor and Actuator Networks. By Andrey V. Savkin, Teddy M. Cheng, Zhiyu Xi, Faizan Javed, Alexey S. Matveev, and Hung Nguyen. Copyright © 2015 by the Institute of Electrical and Electronics Engineers, Inc.

blanket coverage is achieved as the probability of detecting intruders into the region is maximized. Moreover, the network of sensors is in the triangular lattice pattern, except at the boundary of $\mathscr{R}$. The excess sensors form an incomplete layer that lies outside $\mathscr{R}$.

As was already shown in Chapters 6 and 8, the most important feature of deploying sensors in the triangular lattice pattern is that it is optimal in terms of minimum number of sensors required for complete coverage of a bounded set. Moreover, a mobile robotic sensor network in the triangular lattice pattern provides not only coverage but also connectivity, whereas sensing coverage and network connectivity are both important issues in wireless sensor networks [42]. As pointed out in, e.g., [4], given a sensing range r_s and communication range $r_c \geq \sqrt{3}r_s$, deploying sensors in the triangular lattice pattern indeed provides 1-coverage and 6-connectivity. In other words, every point in the region is covered by at least one sensor and also every sensor is connected to six neighboring sensors.

The main contribution of this chapter is a novel approach to blanket coverage of a two-dimensional region between two boundaries by deploying a network of low-power mobile sensors. Each sensor has limited communication capacity, and we theoretically develop a decentralized control algorithm for the network of sensors that is self-deployed and works in a distributed unsupervised mode. In addition, the mobile sensor network forms the triangular lattice pattern.

A potential application of our results is for border protection; see, e.g., [59]. The points L_1 and L_2 can be considered as the two ends of a country's border and the lines $\mathscr{W}_1$ and $\mathscr{W}_2$ can be viewed as the boundaries of the border buffer zone, as is shown in Fig 9.2. The surveillance requirement is that any passing intruder has to be detected between the two ends of the border. To maximize the probability of detecting intruders, we blanket cover the border buffer zone $\mathscr{R}$ with sensors; see Fig 9.2. As a result, any intruder's path that passes $\mathscr{W}_1$ and $\mathscr{W}_2$ in the region $\mathscr{R}$ is detected by a number of distinct sensors; see, e.g., paths A, B, and C in Fig 9.2. One of the advantages of blanket coverage of $\mathscr{R}$, when comparing with barrier coverage as in e.g., [16,59], is that stronger connectivity is achieved; in particular, triangular lattice pattern provides 6-connectivity. As was mentioned above, coverage and connectivity are both important issues in a sensor network. Moreover, our blanket coverage guarantees that any path that is along the region $\mathscr{R}$ (see path D in Fig 9.2) can also be detected by more that one sensor. On the other hand, the barrier coverage in [16,59] cannot guarantee such multi-level detection along $\mathscr{R}$ in the direction of the border. As described in [59], the US–Mexico border spans 2000 miles and it is not feasible to dispatch the sensors at their desired locations in $\mathscr{R}$ on by one. It is therefore natural to consider deploying a mobile sensor network for such a task. By applying our algorithm, we can simply dispatch all the sensors at one end, say point L_1, and let the mobile sensors to relocate themselves in order to triangularly blanket cover $\mathscr{R}$.

Using mobile sensors to achieve blanket coverage has drawn a significant amount of attention in societies such as computer science, communications, and control engineering. The work [29] studies a problem of decentralized coverage control to achieve optimal sensor placement of mobile sensors. It belongs to the area of locational optimization. A control algorithm is proposed in [68] to drive a group of

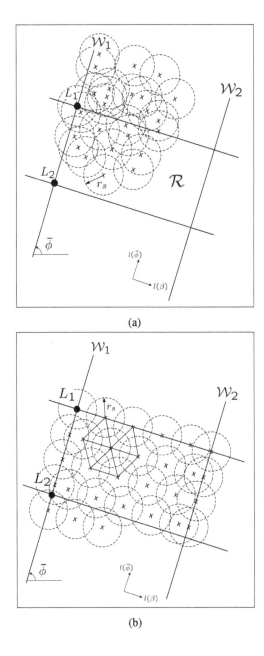

(a)

(b)

Figure 9.1 (a) Initial deployment; (b) Final deployment with triangular blanket coverage in $\mathscr{R}$.

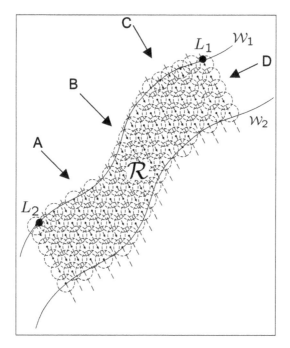

Figure 9.2   Blanket coverage along a belt region that can be considered as the border buffer zone of a country.

mobile sensors to form a square or rectangular lattice pattern. In terms of achieving triangular lattice pattern, a heuristic algorithm is proposed in [66] for a network of mobile sensors, but the algorithm can only achieve a lattice pattern that is close to equilateral triangulation. It should be pointed out that the approach of this chapter is totally different from the approach of Chapter 6. The algorithm developed in this chapter is based on building a one-dimensional structure of mobile sensors and then navigating this structure to solve our blanket coverage problem. Meanwhile, Chapter 6 proposes a distributed randomized control algorithm, which requires finding consensus on a certain regular two-dimensional grid and then finding a suitable location for each mobile node at one of vertices of this grid using a stochastic rule.

As always in this book, in this chapter, each mobile sensor could have severe detection and communication constraints. So a centralized control algorithm is not feasible and the control of such self-deployed mobile sensors falls within the domain of decentralized control. The objective of this chapter is to develop a set of decentralized or distributed control laws for a group of self-deployed mobile robotic sensors to perform triangular blanket coverage for a given region between two boundaries. To the best of our knowledge, this problem has not been studied by other authors. Unlike many other blanket coverage algorithms, the proposed algorithm is asymptotically optimal in terms of minimum number of sensors required for complete coverage of a bounded two-dimensional set.

The main results of this chapter were originally published in [24].

The remainder of the chapter is organized as follows. Section 9.2 formulates the blanket coverage problem studied in this chapter. Sections 9.3 and 9.4 present a randomized algorithm of distributed self-deployment for blanket coverage and its mathematical analysis. Section 9.5 contains computer simulation results illustrating the proposed algorithm. Finally, Section 9.6 presents the proof of the main theoretical result of the chapter.

9.2 Problem of Blanket Coverage between Two Lines

Our objective is to develop a set of control laws to drive a network of autonomous mobile sensors to cover a two-dimensional region $\mathscr{R} \subset \mathbb{R}^2$. To define the region, we first denote $l(s) = \begin{bmatrix} \cos(s) & \sin(s) \end{bmatrix}^T$. Next, we introduce two parallel lines

$$\mathscr{W}_i := \left\{ p \in \mathbb{R}^2 : p^T l(\bar{\phi}) = d_i \right\} \qquad \text{for} \quad i = 1, 2, \tag{9.1}$$

where d_i is a scalar associated with $\mathscr{W}_i$, $l(\bar{\phi})$ is the unit vector that is orthogonal to the line $\mathscr{W}_i$, and $\bar{\phi}$ is the slope of this vector with respect to the x-axis and T denotes the vector transposition. Given two distinct points $L_1, L_2 \in \mathscr{W}_1$, the region $\mathscr{R}$ is then defined by

$$\mathscr{R} := \left\{ p \in \mathbb{R}^2 : p^T l(\beta) \in [L_2^T l(\beta), L_1^T l(\beta)] \text{ and } p^T l(\bar{\phi}) \in [d_1, d_2] \right\}, \tag{9.2}$$

where $\beta := \bar{\phi} - \pi/2$ and $|d_2 - d_1| =: d > 0$. Without any loss of generality, we assume that $\bar{\phi} \in [0, \pi)$.

We consider a mobile robotic sensor network consisting of n number of mobile sensors, labelled 1 through n. Let $v_i(\cdot)$ be the speed of sensor i. The discrete-time kinematic equations of the sensors are given by

$$p_i((k+1)T) = p_i(kT) + Tv_i(kT)l(\theta_i(kT)) \tag{9.3}$$

for $i = 1, 2, \ldots, n$ and $k = 0, 1, 2, \ldots$, where $p_i(\cdot) \in \mathbb{R}^2$ is the pair of the Cartesian coordinates of sensor i and $\theta_i(\cdot) \in \mathbb{R}$ is its heading measured from the x-axis in the counterclockwise direction. The speed v_i and heading θ_i are the control inputs of sensor i. The speed v_i satisfies $|v_i(t)| \leq v_{\max}$ for $i = 1, 2, \ldots, n$ and all $t \geq 0$. The initial heading of each sensor satisfies $\theta_i(0) \in [0, \pi)$ for $i = 1, 2, \ldots, n$. The model (9.3) is a discrete-time model obtained by discretization of the standard first-order linear continuous-time model. Such discrete-time models are very common in the area of multi-sensor systems; see e.g., [19, 21, 92].

Each mobile sensor has a sensing range of $r_s > 0$: It detects objects in a range of $r_s > 0$ for $t \in [kT, (k+1)T)$. Each sensor communicates with their surrounding neighbors in a range of $r_c > 0$ at any discrete time instance $t = kT$ for the coordination of their motions. The constant r_c is called the *communication range*. We assume that $\sqrt{3}r_s \leq r_c$.

Thus, sensor i has the ability to obtain information of its neighbors in a disk of radius r_c defined by

$$D_{i,r_c}(kT) := \{p \in \mathbb{R}^2 : \|p - p_i(kT)\| \leq r_c\},$$

where $\|\cdot\|$ denotes the Euclidean norm. In terms of detection, each sensor can detect or sense any objects in a disk of radius r_s defined by

$$D_{i,r_s}(kT) := \{p \in \mathbb{R}^2 : \|p - p_i(kT)\| \leq r_s\}.$$

Each mobile sensor may have restricted communication and detection capabilities to reduce the cost of operation. As a result, its awareness about the other sensors in the group also may be limited. Therefore, the control law of each mobile sensor should be distributed or decentralized in the sense that the movement of each sensor only relies on information of its neighbors and itself, e.g., positions, headings, and coordination variables. This information is local and limited to its neighbors, as each sensor can only communicate with or observe its neighbors, not all the sensors in the network. For instance, the locations of neighboring sensors can be estimated using recursive state estimation methods; see, e.g., [88, 102, 103].

The communication range r_c of each mobile sensor satisfies $r_c < v_{\max}T/\sqrt{2}$, which can be met by choosing an appropriate sampling period T for a given $v_{\max}$. Moreover, each sensor has the capability of detecting the boundaries $\mathscr{W}_1$ and $\mathscr{W}_2$, as well as the orientations of their tangents. In the case of straight parallel lines $\mathscr{W}_1$ and $\mathscr{W}_2$, both orientations are constant and equal to $\bar{\phi}$.

Let $\mathscr{N}_i(kT)$ be the set of all sensors $j \in \{1, 2, \ldots, n\} \setminus \{i\}$ that at time $t = kT$ belong to the disk $D_{i,r_c}(kT)$, and let $|\mathscr{N}_i(kT)|$ be the number of elements in $\mathscr{N}_i(kT)$. These elements are called *neighbors* of sensor i at time kT. Let $\mathscr{G}$ be the collection of all simple undirected graphs defined on n vertices. For any time $kT \geq 0$, the relation of neighborship is described by a simple undirected graph $G(kT) \in \mathcal{G}$ with the vertex set $\{1, 2, \ldots, n\}$, where i corresponds to sensor i. The vertices $i \neq j$ of the graph $G(kT)$ are connected by an edge if and only if the sensors i and j are neighbors at time kT. The following assumption about the graph sequence is slightly stronger than the standard Main Connectivity Assumption (Assumption 2.4 from Chapter 2).

Assumption 9.1 *The graph $G(kT)$ is connected for all $k \geq 0$.*

Remark 9.1 *In this chapter, we assume that the sensor communication graph $G(kT)$ is undirected. However, in practice, the sensors can have different communication capabilities, which results in a directed communication graph. The case of directed communication graphs was studied in, e.g., [131], where the results of [53, 96] were generalized. It is possible, that by using the approach of [131], the results of this chapter can be extended to the case of directed graphs.*

For a given sensing range r_s, we put

$$K(r_s) := \lceil 2\|L_1 - L_2\|/(3r_s)\rceil + 1, \tag{9.4}$$

where $\lceil x \rceil$ denotes the smallest integer that is greater than or equal to x and $\lfloor x \rfloor$ denotes the largest integer that is smaller than or equal to x. Also, we let

$$n_i(r_s) := \begin{cases} \lfloor d/(\sqrt{3}r_s) \rfloor + 2 & \text{if } i \text{ is odd,} \\ \lfloor (2d - \sqrt{3}r_s)/(2\sqrt{3}r_s) \rfloor + 3 & \text{if } i \text{ is even} \end{cases} \tag{9.5}$$

for $i = 1, 2, \ldots, K(r_s)$.

By using $K(r_s)$, $n_1(r_s)$, and $n_2(r_s)$, we denote by $m(r_s)$ the following integer

$$m(r_s) = \begin{cases} \dfrac{K(r_s) + 1)n_1(r_s) + (K(r_s) - 1)n_2(r_s)}{2} & \text{if } K(\cdot) \text{ is odd,} \\ \dfrac{K(r_s)n_1(r_s) + K(r_s)n_2(r_s)}{2} & \text{if } K(\cdot) \text{ is even.} \end{cases} \tag{9.6}$$

Here $m(r_s)$ is the number of sensors required to fully cover $\mathcal{R}$ with the triangular lattice pattern. Next, we define $m(r_s)$ number of desired locations of the sensors that ensure coverage of $\mathcal{R}$. These $m(r_s)$ locations are *unknown* to the sensors. Before doing this, we define $K(r_s)$ number of points on the line $\mathcal{W}_1$ as follows:

$$a_1 := L_1, \quad a_i := a_1 - (i - 1)r_s \frac{3}{2}l(\bar{\phi}), \tag{9.7}$$

where $i = 2, 3, \ldots, K(r_s)$. Then, using a_i, the desired sensor locations $h_{i,j}(r_s)$ are defined. Specifically if i is odd,

$$h_{i,j}(r_s) = \begin{cases} a_i + \sqrt{3}r_s(j - 1)l(\beta) & \text{if } j = 1, 2, \ldots, n_i(r_s) - 1, \\ a_i + dl(\beta) & \text{if } j = n_i(r_s). \end{cases} \tag{9.8}$$

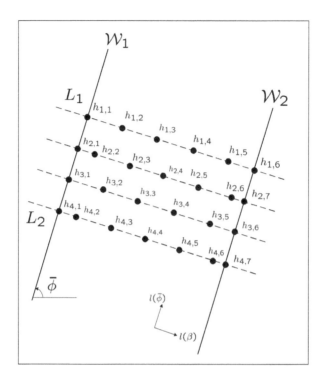

Figure 9.3   Triangular blanket coverage ($K = 4$, $m = 26$) of the region $\mathscr{R}$.

If i is even, then we have

$$h_{i,j}(r_s) = \begin{cases} a_i & \text{if } j = 1, \\ a_i + \sqrt{3}r_s(j - \dfrac{3}{2})l(\beta) & \text{if } j = 2, 3 \ldots, n_i(r_s) - 1, \\ a_i + dl(\beta) & \text{if } j = n_i(r_s). \end{cases} \qquad (9.9)$$

From their definitions, the locations a_i, b_i, and the desired sensor location $h_{i,j}$ are all dependent upon the sensing range r_s. In other words, different values of r_s give rise to different sets of desired sensor locations. No matter what the value r_s is, the corresponding desired sensor locations will always form the triangular lattice pattern.

Definition 9.1 *Let $m(r_s)$ be the number of sensors with sensing range r_s that cover a bounded region $\mathscr{R} \subset \mathbb{R}^2$, and let $\mathscr{P}(r_s) = \{h_{i,j}(r_s)\}$ be the corresponding set of sensor locations in $\mathscr{R}$. Also let $N(r_s)$ be the minimum number of sensors with sensing range r_s that can cover $\mathscr{R}$. A family of coverages $\mathscr{P}(r_s)$ with parameter r_s is said to be* asymptotically optimal *for covering $\mathscr{R}$ if for all $p \in \mathscr{R}$, there exists $h_{i,j}(r_s) \in \mathscr{P}(r_s)$ such that*

$$\|p - h_{i,j}(r_s)\| \le r_s, \qquad (9.10)$$

and

$$\lim_{r_s \to 0} r_s^2 m(r_s) = \lim_{r_s \to 0} r_s^2 N(r_s). \tag{9.11}$$

Like similar theorems from Chapters 6 and 8, the following fact easily follows from the result of Kershner [56].

Theorem 9.1 *Let $\mathscr{R} \subset \mathbb{R}^2$ be the given region between the lines $\mathscr{W}_1$ and $\mathscr{W}_2$. Then the family of coverages by the set of sensor locations $\{h_{i,j}(r_s)\}$ defined in (9.8), (9.9) is asymptotically optimal for covering $\mathscr{R}$.*

Proof: For a given r_s, it is immediate from (9.8), (9.9) that $h_{i,j}(r_s)$, $i = 1, 2, \ldots, K(r_s)$, $j = 1, 2, \ldots, n_i(r_s)$ form the triangular lattice pattern in $\mathscr{R}$ and the edge of each tri-angle has a length of $\sqrt{3}r_s$, except at the boundary of $\mathscr{R}$. The sensors placed at the boundaries are to guarantee that every point in the boundaries of $\mathscr{R}$ is also covered. So it is obvious that blanket coverage is achieved by placing the sensors at $h_{i,j}(r_s)$. The condition (9.11) can be directly obtained from [56]. In fact, as is shown in [56], $\lim_{r_s \to 0} \pi r_s^2 N(r_s) = \lim_{r_s \to 0} \pi r_s^2 m(r_s) = (2\pi\sqrt{3}/9)A(\mathscr{R})$, where $A(\mathscr{R})$ is the area of $\mathscr{R}$. This completes the proof of Theorem 9.1.   ∎

Theorem 9.1 states that placing sensors at $h_{i,j}(r_s)$ provides complete coverage for $\mathscr{R}$. In other words, any point in $\mathscr{R}$ is covered by at least one sensor. In addition, it also states that the triangular lattice pattern of $h_{i,j}(r_s)$ is asymptotically optimal in terms of minimum number of sensors required to blanket cover $\mathscr{R}$.

Definition 9.2 (Triangular blanket coverage) *Given a region $\mathscr{R}$ and n mobile sensors, a set of distributed control laws is called the* triangular blanket coverage control *for the network of mobile sensors for covering $\mathscr{R}$ if for almost all initial sensor positions and for each location $h_{i,j}(r_s)$ with $i \in \{1, 2, \ldots, K(r_s)\}$ and $j \in \{1, 2, \ldots, n_i(r_s)\}$, there exists a unique index $z \in \{1, 2, \ldots, n\}$ such that*

$$\lim_{k \to \infty} \| p_z(kT) - h_{i,j}(r_s) \| = 0. \tag{9.12}$$

In Definition 9.2, "almost all" means "all except for a set of zero Lebesgue measure". Provided that there are sufficiently many sensors in the network, i.e., $n \geq m(r_s)$, the region $\mathscr{R}$ is covered by a network of sensors in the triangular lattice pattern, as is shown in Fig. 9.1(b). Excess sensors form extra layers of sensor arrays that lie outside $\mathscr{R}$.

9.3 Blanket Coverage Algorithm

In this section, a control algorithm is offered for the coordination of the mobile robotic sensors, or simply sensors, to achieve triangular blanket coverage for $\mathscr{R}$. We first describe the intuitive idea behind the algorithm and then present its mathe-matical development.

9.3.1 Description

The idea of the proposed control algorithm is to drive a network of mobile robotic sensors to form a number of parallel sensor arrays between the lines $\mathcal{W}_1$ and $\mathcal{W}_2$; see Fig. 9.4(a). Initially, the sensors are dispatched near the point L_1. By applying the algorithm to the sensors, they spread along the line $\mathcal{L}_1$ until this line of sensors reaches $\mathcal{W}_2$. Once this line of sensors reaches $\mathcal{W}_2$, it turns around and continues to form another array of sensors along the line $\mathcal{L}_2$. Similarly, as this second line of sensors reaches $\mathcal{W}_1$, it again turns around and continues to form an array of sensors along the line $\mathcal{L}_3$. The sensors continue to form a sensor lattice layer by layer until the sensors cease spreading. To form the required triangular structure, as is shown in Fig. 9.4(b), sensor i obtains the locations of sensors f, g, and h, and this information is then utilized to determine where to place sensor i. This happens at the locations A, B, and C, as is shown in Fig. 9.4(b), in order to form the interlacing pattern.

9.3.2 Control Laws

Now a set of decentralized control laws is developed in details for the coordination of the sensors to achieve triangular blanket coverage for the case where the region is defined by the parallel lines $\mathcal{W}_1$ and $\mathcal{W}_2$. Since the control laws for the mobile sensors are distributed or decentralized, they rely on the local information of each sensor. Information such as locations and coordination variables of the sensor's neighbors is available to the sensor. One of the coordination variables is $\phi_i(\cdot)$ for $i = 1, 2, \ldots, n$. The variable $\phi_i(\cdot)$ is available to any neighboring sensors of sensor i and is used for coordination of the motion of sensor i with the other sensors in the group. The coordination variable ϕ_i is initialized as

$$\phi_i(0) = \theta_i(0), \quad i = 2, 3, \ldots, n.$$

For sensor 1, its coordination variable $\phi_1(\cdot)$ is constant and given by $\phi_1(\cdot) \equiv \bar{\phi}$, where $\bar{\phi} = \beta + \pi/2$. Also, sensor 1 is placed at L_1 to indicate where the blanket coverage should start. In addition, we introduce the variable $r_i(kT) \in \{1, 2, 3 \ldots\}$ for $i = 1, 2, \ldots, n$, with $r_i(0) = 1$ and $r_1(\cdot) \equiv 1$. This variable characterizes the row that sensor i belongs to.

At time kT, $k = 0, 1, 2, \ldots$, we first put $\mathscr{S}_i(kT) = \{j \in \mathscr{N}_i(kT) : r_j(kT) = r_i(kT)\}$. The set $\mathscr{S}_i(kT)$ contains the neighbors of sensor i that belong to the same row as sensor i. For sensor i, we define the average of the coordination variables from the set $\mathscr{S}_i(kT)$ as follows:

$$\mathscr{H}_i(kT) := \frac{1}{1 + |\mathscr{S}_i(kT)|} \left(\phi_i(kT) + \sum_{j \in \mathscr{S}_i(kT)} \phi_j(kT) \right) \tag{9.13}$$

for $i = 2, 3, \ldots, n$, where $|\mathscr{S}_i(kT)|$ denotes the number of elements in the set $\mathscr{S}_i(kT)$. The coordination variable $\phi_i(kT)$ is updated by

$$\phi_i((k+1)T) = \mathscr{H}_i(kT). \tag{9.14}$$

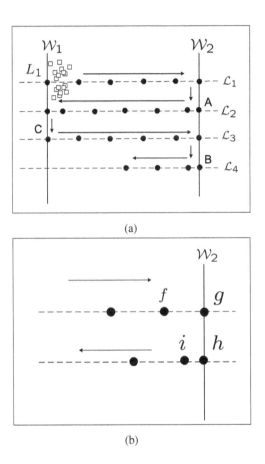

(a)

(b)

Figure 9.4   (a) Forming a triangular lattice (□ initial position, ● sensor position). (b) Sensor placement at the turn around a marginal point.

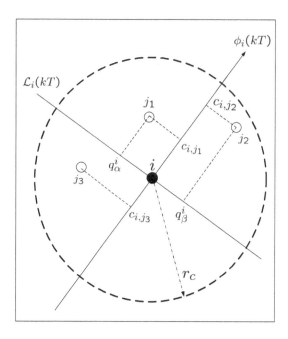

Figure 9.5   Sensor i (•) with neighboring sensors j_1, j_2, and j_3 (○).

For each neighbor of sensor i, we introduce the variable

$$c_{i,j}(kT) = l(\phi_i(kT))^T p_j(kT) \tag{9.15}$$

for $j \in \mathcal{S}_i(kT) \cup \{i\}$. This variable is the projection of $p_j(kT)$ onto the direction $\phi_i(kT)$, as is shown in Fig. 9.5.

Similar to (9.13), we define the average of $c_{i,j}(\cdot)$, $j \in \mathcal{S}_i(kT) \cup \{i\}$ for sensor i:

$$\mathcal{M}_i(kT) := \frac{1}{1 + |\mathcal{S}_i(kT)|} \left(c_{i,i}(kT) + \sum_{j \in \mathcal{S}_i(kT)} c_{i,j}(kT) \right). \tag{9.16}$$

For each sensor i, we introduce another coordination variable $\mathcal{F}_i(kT) := c_{i,i}(kT)$. The variables $\phi_i(\cdot)$ and $\mathcal{F}_i(\cdot)$ are for coordination with other sensors in the group. Using $\phi_i(kT)$ and $\mathcal{F}_i(kT)$, we define a line $\mathcal{L}_i(kT)$ for sensor i as follows:

$$\mathcal{L}_i(kT) = \{p \in \mathbb{R}^2 : l(\phi_i(kT))^T p = \mathcal{F}_i(kT)\} \tag{9.17}$$

for $i = 1, 2, \ldots, n$ and $k = 0, 1, 2, \ldots$. This line is shown in Fig. 9.5 and is used for defining the control for sensor i.

To develop the control action along $\mathcal{L}_i(kT)$, we denote by $q^i_j(kT)$ the projection of the position of sensor $j \in \mathcal{S}_i(kT) \cup \{i\}$ on the line $\mathcal{L}_i(kT)$ at time kT. It is given by

$q_j^i(kT) = l(\beta_i(kT))^T p_j(kT)$, where $\beta_i(kT) := \phi_i(kT) - \pi/2$. Using this, we define $\alpha, \beta \in \mathscr{S}_i(kT)$, provided that if they exist, such that

$$q_\alpha^i(kT) < q_i^i(kT) < q_\beta^i(kT) \tag{9.18}$$

and $q_\alpha^i(kT)$ and $q_\beta^i(kT)$ are the left and right most closest to $q_i^i(kT)$, respectively. For example, as is shown in Fig. 9.5, $q_\alpha^i(kT)$ and $q_\beta^i(kT)$ are equal to $q_{j_1}^i(kT)$ and $q_{j_2}^i(kT)$, respectively.

First, we consider the case where $r_i(kT)$ is **odd**. If both sensors α and β exist, and $p_\alpha(kT) \notin \mathscr{W}_1$ or $p_\beta(kT) \notin \mathscr{W}_2$, then we put

$$\mathscr{Q}_i(kT) = \frac{q_\alpha^i(kT) + q_\beta^i(kT)}{2}. \tag{9.19}$$

If α exists but not β, or if α exists and β is on $\mathscr{W}_2$, then we define

$$\mathscr{Q}_i(kT) = \frac{q_\alpha^i(kT) + q_i^i(kT) + s}{2}, \tag{9.20}$$

where $s := \sqrt{3}r_s$. On the other hand, if β exists but not α, then we define

$$\mathscr{Q}_i(kT) = \frac{q_\beta^i(kT) + q_i^i(kT) - s}{2}. \tag{9.21}$$

If sensor i hits $\mathscr{W}_2$ and there are no other sensors in $\mathscr{S}_i(kT)$ that are on $\mathscr{W}_2$, sensor i is placed at $\mathscr{W}_2$ and we put

$$\mathscr{Q}_i(kT) = q_{\eta,2}^i(kT), \tag{9.22}$$

where $\eta_{i,2}(kT) := \mathscr{L}_i(kT) \cap W_2$ and $q_{\eta,2}^i(kT) = l(\phi_i(kT) - \pi/2)^T \eta_{i,2}(kT)$. When sensor i moves along $\mathscr{W}_1$ from row $r_i(kT) - 1$ and sensors α and β do not exist, then sensor i is placed at $\mathscr{W}_1$ and we define

$$\mathscr{Q}_i(kT) = q_{\eta,1}^i(kT), \tag{9.23}$$

where $\eta_{i,1}(kT) := \mathscr{L}_i(kT) \cap \mathscr{W}_1$ and $q_{\eta,1}^i(kT) = l(\phi_i(kT) - \pi/2)^T \eta_{i,1}(kT)$.

As discussed before, if sensor i has its neighbor sensor α at $\mathscr{W}_1$ and there are no sensors between sensors i and α, i.e., sensor i is the second sensor from left in the row $r_i(kT)$, then sensor i needs to know the positions of sensors α, γ, and δ (see Fig. 9.6(a)) to determine where it should be placed in order to create the interlacing feature of the triangular lattice pattern. We let $\rho_i(kT)$ be the required distance from sensor α to sensor i at time kT. Then we define

$$\mathscr{Q}_i(kT) = \frac{q_\alpha^i(kT) + q_i^i(kT) + \rho_i(kT)}{2}. \tag{9.24}$$

Next, we consider the case where $r_i(kT)$ is **even**. If both sensors α and β exist, and $p_\alpha(kT) \notin \mathscr{W}_1$ or $p_\beta(kT) \notin \mathscr{W}_2$, then we define

$$\mathscr{Q}_i(kT) = \frac{q_\alpha^i(kT) + q_\beta^i(kT)}{2}. \tag{9.25}$$

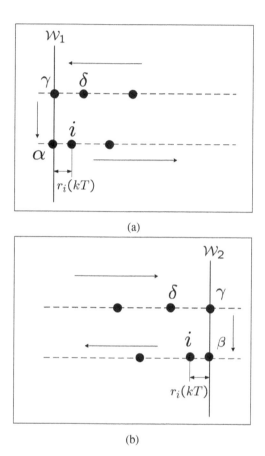

(a)

(b)

Figure 9.6   Determination of the desired sensor i location: (a) Odd r_i; (b) Even r_i.

If β exists but not α, or if β exists and α is on $\mathscr{W}_1$, then we define

$$\mathscr{D}_i(kT) = \frac{q_\beta^i(kT) + q_i^i(kT) - s}{2}. \qquad (9.26)$$

If α exists but not β, then we define

$$\mathscr{D}_i(kT) = \frac{q_\alpha^i(kT) + q_i^i(kT) + s}{2}. \qquad (9.27)$$

If sensor i hits $\mathscr{W}_1$ and no other sensor is on $\mathscr{W}_1$, sensor i is placed at $\mathscr{W}_1$ and we define

$$\mathscr{D}_i(kT) = q_{\eta,1}^i(kT). \qquad (9.28)$$

When sensor i moves along $\mathscr{W}_2$ from line $r_i(kT) - 1$ and sensors α and β do not exist, then sensor i is placed at $\mathscr{W}_2$ and we define

$$\mathscr{D}_i(kT) = q_{\eta,2}^i(kT). \qquad (9.29)$$

Similar to the case where $r_i(kT)$ is odd, if sensor i has its neighbor sensor β at $\mathscr{W}_2$ and there are no sensors between sensors i and β, then sensor i needs to use the positions of sensors β, γ, and δ (see Fig. 9.6(b)) to determine $\rho_i(kT)$ in order to produce the triangular structure. Then we define

$$\mathscr{D}_i(kT) = \frac{q_\beta^i(kT) + q_i^i(kT) - \rho_i(kT)}{2}. \qquad (9.30)$$

Using $\mathscr{D}_i(kT)$, we define the velocity component that is along the line $\mathscr{L}_i(kT)$ as follows:

$$\bar{v}_i(kT) = \begin{cases} 0 & \text{if } \mathscr{D}_i(kT) \leq q_{\eta,1}^i(kT) \\ & \text{or } \mathscr{D}_i(kT) \geq q_{\eta,2}^i(kT), \\ \dfrac{\mathscr{D}_i(kT) - q_i^i(kT)}{T} & \text{otherwise.} \end{cases} \qquad (9.31)$$

To determine the heading $\theta_i(kT)$ of sensor i, we also put

$$\chi_i(kT) = \begin{cases} \phi_i(kT) & \text{if } q_{\eta,1}^i(kT) < \mathscr{D}_i(kT) < q_{\eta,2}^i(kT), \\ \psi_i(kT) & \text{otherwise,} \end{cases} \qquad (9.32)$$

where $\psi_i(kT)$ is the gradient of $\mathscr{W}_1$ or $\mathscr{W}_2$ detected by sensor i at time kT. The gradients of $\mathscr{W}_1$ or $\mathscr{W}_2$ are unknown to the sensors, as the sensors have no prior information of the environment where they are deployed. To determine $\psi_i(kT)$, each sensor is required to have the capability of detecting the slope of $\mathscr{W}_1$ or $\mathscr{W}_2$. For the case with parallel straight lines $\mathscr{W}_1$ and $\mathscr{W}_2$, $\psi_i(kT) \equiv \bar{\phi}$ for all $k \geq 0$.

If sensor i is placed at $\mathscr{W}_1$ when $r_i(kT)$ is odd, or at $\mathscr{W}_2$ when $r_i(kT)$ is even, then we define $\mathscr{F}_i(kT)$, γ_i, and $\bar{\mathscr{M}}_i(kT)$ as follows

$$\mathscr{F}_i(kT) := \{j \in \mathscr{N}_i(kT) \mid r_j(kT) = r_i(kT) - 1\},$$

$$\gamma_i := \arg\min_{j \in \mathscr{F}_i(kT)} |l(\phi_i(kT))^T (p_j(kT) - p_i(kT))|, \qquad (9.33)$$

$$\bar{\mathscr{M}}_i(kT) := \left(c_{i,\gamma_i}(kT) + c_{i,i}(kT) - s\sqrt{3}/2\right)/2.$$

The value $\bar{\mathscr{M}}_i(kT)$ is for maintaining the distance between two rows of sensor arrays at $s(\sqrt{3}/2)$. Using $\mathscr{M}_i(kT)$, $\bar{\mathscr{M}}_i(kT)$, and $\mathscr{F}_i(kT)$, we introduce

$$\mathscr{M}_i(kT) = \begin{cases} \mathscr{F}_i(kT) - s\sqrt{3}/2 & \text{if } \mathscr{Q}_i(kT) \leq q_{\eta,1}^i(kT), \\ & \text{or } \mathscr{Q}_i(kT) \geq q_{\eta,2}^i(kT), \\ \bar{\mathscr{M}}_i(kT), & \text{if } p_i(kT) \in \mathscr{W}_1 \text{ or } \mathscr{W}_2, \\ \mathscr{M}_i(kT), & \text{otherwise.} \end{cases} \qquad (9.34)$$

To update the row number, we use the following rule

$$r_i((k+1)T) = \begin{cases} r_i(kT) & \text{if } q_{\eta,1}^i(kT) < \mathscr{Q}_i(kT) < q_{\eta,2}^i(kT), \\ r_i(kT) + 1 & \text{otherwise,} \end{cases} \qquad (9.35)$$

with $r_i(0) = 1$.

Before introducing our decentralized control laws, we define

$$\hat{v}_i(kT) = (\mathscr{M}_i(kT) - \mathscr{F}_i(kT))/T \qquad (9.36)$$

for $i = 2, 3, \ldots, n$ and $k = 0, 1, 2, \ldots$, which is the desired velocity of sensor i orthogonal to the line $\mathscr{L}_i(kT)$. Using (9.31) and (9.36), we introduce a set of decentralized control laws that is described by:

$$v_i(kT) = \sqrt{\bar{v}_i(kT)^2 + \hat{v}_i(kT)^2};$$

$$\theta_i(kT) = \begin{cases} \chi_i(kT) + \zeta_i(kT) - \pi/2 & \text{if } \hat{v}_i(kT) \geq 0, \\ \chi_i(kT) - \zeta_i(kT) - \pi/2 & \text{if } \hat{v}_i(kT) < 0, \end{cases} \qquad (9.37)$$

where $\zeta_i(kT) := \cos^{-1}(\bar{v}_i(kT)/v_i(kT))$, for $i = 1, 2, \ldots, n$ and $k = 0, 1, 2, \ldots$.

9.3.3 Algorithm Convergence

Theorem 9.2 *Consider n mobile sensors with sensing range r_s and also consider a region $\mathscr{R}$ defined by (9.2). Suppose that Assumption 9.1 holds and $n \geq m(r_s)$. Then the set of decentralized control laws (9.37) is a triangular blanket coverage control for the network of mobile sensors for covering $\mathscr{R}$.*

The proof of Theorem 9.2 is given in Section 9.6.

Remark 9.2 *In practice, a mobility range of sensors is an important issue. It follows from the proof of Theorem 9.2 (see Section 9.6) that the maximum distance traveled by all sensors under the algorithm (9.37) is less than $\sqrt{3}r_s n$ for small enough r_s.*

9.4 Triangular Blanket Coverage between Curves

In this section, our triangular blanket coverage problem is extended to a two-dimensional region that is bounded by two smooth curves, instead of straight parallel lines. Let $\mathscr{W}_1$ and $\mathscr{W}_2$ be two smooth and non self-intersecting curve segments lying in $\mathscr{D} \subset \mathbb{R}^2$. We assume that for each curve segment $\mathscr{W}_i$, there exist $\bar{\phi}_i \in [0, \pi)$, $d_{i,1}$, and $d_{i,2}$ such that

$$\mathscr{W}_i \subset \{p \in \mathscr{D} : p^T l(\bar{\beta}_i) \in [d_{i,1}, d_{i,2}]\}, \tag{9.38}$$

where $\bar{\beta}_i := \bar{\phi}_i - \pi/2$. Each $\mathscr{W}_i$ lies between two parallel lines, namely $\{p \in \mathscr{D} : p^T l(\bar{\beta}_i) = d_{i,1}\}$ and $\{p \in \mathscr{D} : p^T l(\bar{\beta}_i) = d_{i,2}\}$.

Next, let $\mathscr{B} \subset \mathscr{D}$ be the region between $\mathscr{W}_1$ and $\mathscr{W}_2$. Given $L_1, L_2 \in \mathscr{W}_1$, and $\bar{\phi} \in [\min(\bar{\phi}_1, \bar{\phi}_2), \max(\bar{\phi}_1, \bar{\phi}_2)]$, the region $\mathscr{R}$ that is required to be triangularly blanket covered is defined by

$$\mathscr{R} := \{p \in \mathscr{D} : L_1^T l(\bar{\phi}) \geq p^T l(\bar{\phi}) \geq L_2^T l(\bar{\phi})\} \cap \mathscr{B}, \tag{9.39}$$

as is shown in Fig. 9.7(a).

Similar to the straight lines case, we introduce desired sensor locations in $\mathscr{R}$. To this end, we first put

$$K(r_s) := \lceil 2(L_1 - L_2)^T l(\bar{\phi})/(3r_s) \rceil + 1. \tag{9.40}$$

Then $K(r_s)$ parallel lines are defined as follows:

$$\mathscr{L}_i := \left\{ p \in \mathscr{D} : p^T l(\bar{\phi}) = L_1^T l(\bar{\phi}) + r_s(i-1)\frac{3}{2} \right\}. \tag{9.41}$$

Using these lines, we introduce the following points on $\mathscr{W}_1$ and $\mathscr{W}_2$:

$$a_i := \mathscr{L}_i \cap \mathscr{W}_1, \quad b_i := \mathscr{L}_i \cap \mathscr{W}_2 \tag{9.42}$$

for $i = 1, 2, \ldots, K(r_s)$. For each $i = 1, 2, \ldots, K(r_s)$, let $n_i(r_s)$ and $\bar{n}_i(r_s)$ be the number and maximum number of sensors between the points a_i and b_i, respectively. Then, for $r_s < \|a_i - b_i\|$, we have

$$n_i \leq \bar{n}_i = \lceil \|2(a_i - b_i)/(3r_s)\| \rceil + 2. \tag{9.43}$$

We define $m(r_s) := \sum_i^K n_i(r_s)$ as the total number of desired sensor locations and $\bar{n} := \sum_i^K \bar{n}_i$ as its upper bound. By using a similar approach as in (9.8), (9.9), each desired sensor location $h_{i,j}(r_s)$ can be derived for $i = 1, 2, \ldots, K(r_s)$ and $j = 1, 2, \ldots, n_i(r_s)$, and these locations are shown in Fig. 9.7(b). We denote the set of these locations by

$$\Gamma(r_s) := \{h_{i,j}(r_s)\}. \tag{9.44}$$

Again, these locations are unknown to the sensors.

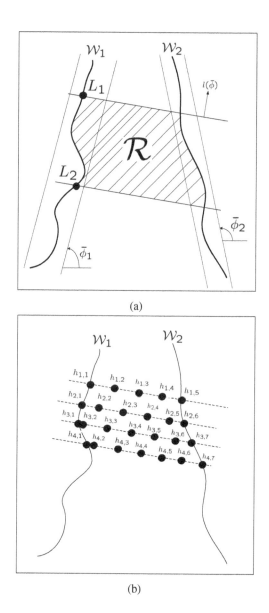

Figure 9.7   (a) The region $\mathcal{R}$ to be covered. (b) Desired sensor locations.

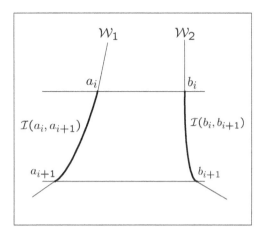

Figure 9.8   The length of the curve segment between a_i and a_{i+1}, or b_i and b_{i+1}.

Assumption 9.2 *For any $i = 1, 2, \ldots, K - 1$,*

$$\|a_i - a_{i+1}\| \le \sqrt{3}r_s \quad and \quad \|b_i - b_{i+1}\| \le \sqrt{3}r_s. \tag{9.45}$$

This assumption is introduced to ensure that sensors placed at a_i and a_{i+1}, or b_i and b_{i+1} are connected, i.e., the distance between them is no greater than r_c. Also, we denote by $\mathcal{I}(p_1, p_2)$ the length of a smooth curved segment with the end points p_1 and $p_2 \in \mathcal{W}_1$ or $\mathcal{W}_2$.

Assumption 9.3 *For any $i = 1, 2, \ldots, K - 1$,*

$$\mathcal{I}(a_i, a_{i+1}) \le 2r_s \quad and \quad \mathcal{I}(b_i, b_{i+1}) \le 2r_s. \tag{9.46}$$

Assumption 9.3 guarantees that any point along the curve segment of $\mathcal{W}_1$ between points a_i and a_{i+1} is covered by at least one of the sensors located at a_i and a_{i+1}; see Fig. 9.8. Similarly, it guarantees that any point along the curve segment of $\mathcal{W}_2$ between points b_i and b_{i+1} is covered by at least one of the sensors located at b_i and b_{i+1}. In other words, Assumption 9.3 is introduced to ensure that any point along the boundaries between the lines $\mathcal{L}_i$ and $\mathcal{L}_{i+1}$ is covered by at least one sensor.

Theorem 9.3 *Let $\mathcal{R} \subset \mathbb{R}^2$ be the region between the curves $\mathcal{W}_1$ and $\mathcal{W}_2$. Then the family of coverages by the set of sensor locations $\Gamma(r_s)$ is asymptotically optimal for covering $\mathcal{R}$.*

The proof of this theorem is similar to the proof of Theorem 9.1.
     We are now in a position to state the main result of this chapter.

Theorem 9.4 *Consider n mobile sensors with sensing range r_s, and also consider the region $\mathcal{R}$ defined by (9.39). Suppose that Assumptions 9.1—9.3 hold, and $n \ge$*

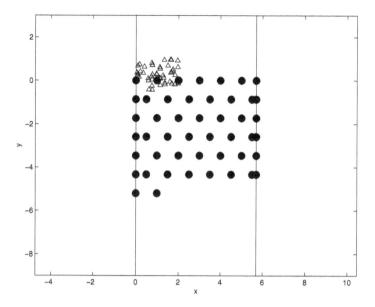

Figure 9.9 Triangular blanket coverage ($\bar{\phi} = \pi/2, n = 47, r_s = 1/\sqrt{3}$): sensor positions ($\triangle$ initial, • final).

$m(r_s)$. *Then the set of decentralized control laws (9.37) is a triangular blanket coverage control for the network of mobile sensors for covering $\mathcal{R}$.*

The proof of this theorem is similar to the proof of Theorem 9.2.

9.5 Illustrative Examples

In this section, we present simulation results to illustrate the proposed algorithm. In the first set of simulations, our objective is to obtain triangular blanket coverage within the region $\mathcal{R}$ defined by the points $(0,0), (0,-4), (5.7,0)$, and $(5.7,-4)$. The points L_1, L_2 are $(0,0)$ and $(0,-4)$, respectively. As is shown in Fig. 9.9, the sensors are initially scattered around the point $(0,0)$. After applying the proposed algorithm, the sensors are in formation that provides blanket coverage in $\mathcal{R}$. Moreover, the formation exhibits the triangular lattice pattern, except for a vicinity of the boundaries of $\mathcal{R}$. Also, the excess sensors form an incomplete layer that lies outside $\mathcal{R}$.

Figure 9.10 illustrates the evolution of the sensor positions. They are initially dispatched near L_1, as is indicated in Fig. 9.10(a). Figure 9.10(b) shows that at time $kT = 40$ the sensors are aligned, forming the first row of sensor array along $\mathcal{L}_1$. There is a sensor already moved down by $3r_s/2$, where the second row of sensor array should be formed. At time $kT = 150$, the second row of sensor array is formed, as is shown in Fig. 9.10(c). As time proceeds (see Figs. 9.10(d), (e)), the snake of sensors continues to spread and more rows of sensor arrays are formed. In Fig. 9.10(f), the network of sensors forms a sensor lattice at time $kT = 1300$, which is in the triangular lattice pattern between the given points L_1 (▼) and L_2 (▲).

For $\bar{\phi} = \pi/4$, $n = 33$, and the region $\mathcal{R}$ defined by the points $(0,0)$, $(4.03, -4.03)$, $(-1.84, -1.84)$, and $(2.2, -5.87)$, similar results are also obtained. The points L_1 and L_2 are given by $(0,0)$ and $(-1.84, -1.84)$, respectively. In terms of blanket coverage for a region between two arbitrary curves, our proposed algorithm drives the sensor network into a triangular sensor lattice pattern in both cases, as is shown in Figs. 9.12 and 9.13.

As discussed in Section 9.1, one of the potential applications of our proposed algorithm is for border protection. To illustrate this application, suppose we wish to blanket cover a belt region, which may be viewed as the border buffer zone of a country and the two points L_1 and L_2 are defined as the end points of the border; see Fig. 9.14. For illustration, we deployed $n = 180$ mobile sensors to blanket cover the border buffer zone. Using our proposed algorithm, the zone is fully covered by the sensors, as is shown in Figure 9.15, and most importantly, the sensors are in the triangular lattice pattern.

9.6   Proof of Theorem 9.2

Proof of Theorem 9.2 By the connectivity assumption, the sensors are connected to the line $\mathcal{W}_1$ via sensor 1, which is placed at L_1. Using the results in, e.g., [53, 92], Assumption 9.1, and the update laws for ϕ_i and $\mathcal{F}_i(kT)$, it is easy to see that the sensors eventually converge to the line $\mathcal{L}_1$ and form an array of sensors. This array of sensors spreads and turns around when it hits $\mathcal{W}_2$, as is shown in Fig. 9.4(a). Then it forms the second array of sensors along $\mathcal{L}_2$. By doing so, this snake of sensors forms multiple layers of parallel sensor arrays. The control laws not only keep the sensors aligned along each layer of sensor array, but also drive the distance between neighboring sensors to s, except for the sensors placed at $\mathcal{W}_1$ and $\mathcal{W}_2$. Therefore, the snake of sensors eventually reaches a position where it ceases spreading.

To justify the above argument, we will act layer by layer. For the first layer, sensor i has the initial state $\{p_i(0), \theta_i(0)\}$ for $i = 2, 3, \ldots, n$, with $p_1(kT) \equiv L_1$ and $\theta_1(kT) = \bar{\phi}$ for all $k \geq 0$. Suppose that the line $\mathcal{W}_2$ is absent. The connectivity property imposed in Assumption 9.1 and the update law for ϕ_i guarantee that

$$\lim_{k \to \infty} \phi_i(kT) = \bar{\phi}, \quad i = 2, 3, \ldots, n. \tag{9.47}$$

That is, sensors $1, 2, \ldots, n$, reach a consensus in terms of the coordination variable $\phi_i(\cdot)$ with sensor 1 as the leader; see, e.g., [53]. Similarly, since $\phi_i(kT) \to \bar{\phi}$ and $\mathcal{M}_i(kT) \to \bar{\mathcal{F}}_1 := l(\bar{\phi})^T L_1$, we have

$$\lim_{k \to \infty} \mathcal{F}_i(kT) = \bar{\mathcal{F}}_1, \quad i = 2, 3, \ldots, n. \tag{9.48}$$

In other words, the conditions (9.47) and (9.48) guarantee that there exists a line

$$\mathcal{L}_1 = \{p \in \mathbb{R}^2 : l(\bar{\phi})^T p = \bar{\mathcal{F}}_1\} \tag{9.49}$$

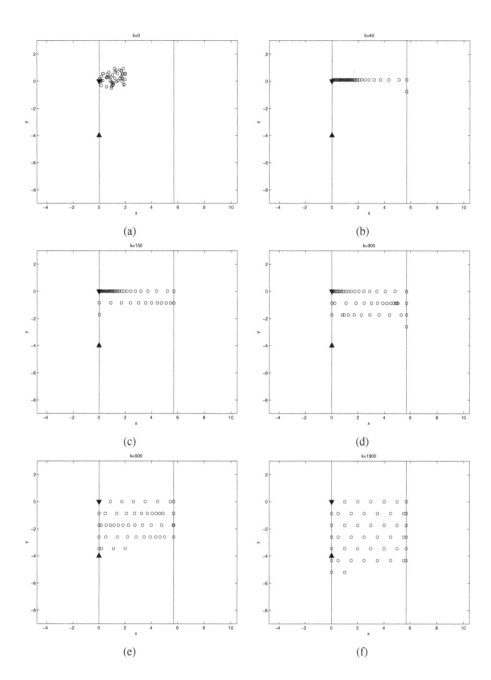

Figure 9.10   Forming a triangular lattice pattern ($\blacktriangledown$-L_1 , $\blacktriangle$-L_2, $\circ$-sensor position): (a) $kT = 0$; (b) $kT = 40$; (c) $kT = 150$; (d) $kT = 300$; (e) $kT = 500$; (f) $kT = 1300$ ($n = 47$, $r_s = 1/\sqrt{3}$).

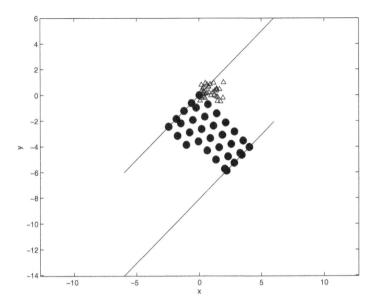

Figure 9.11   Triangular blanket coverage ($\bar{\phi} = \pi/4$, $n = 33$, $r_s = 1/\sqrt{3}$): sensor positions ($\triangle$-initial, •-final).

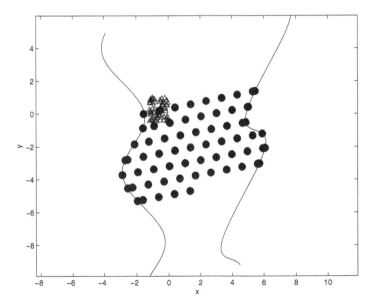

Figure 9.12   Triangular blanket coverage with non-straight $\mathcal{W}_1$ and $\mathcal{W}_2$ ($n = 64$, $r_s = 1/\sqrt{3}$): sensor positions ($\triangle$-initial, •-final).

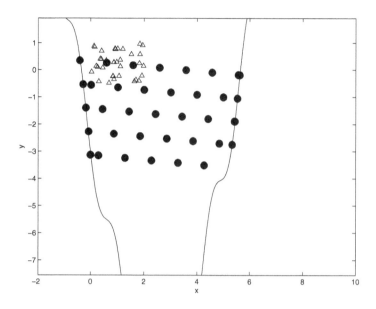

Figure 9.13   Triangular blanket coverage with non-straight $\mathscr{W}_1$ and $\mathscr{W}_2$ ($n = 36$, $r_s = 1/\sqrt{3}$): sensor positions ($\triangle$ initial, $\bullet$ final).

such that $L_1, p_1 \in \mathscr{L}_1$ and

$$\lim_{k \to \infty} d(p_i(kT), \mathscr{L}_1) = 0, \quad i = 2, 3, \ldots, n, \tag{9.50}$$

where $d(p_i(\cdot), \mathscr{L}_1)$ is the distance between the point $p_i(\cdot)$ and the line $\mathscr{L}_1$. Thus, all the sensors $i = 2, 3, \ldots, n$ converge to the line $\mathscr{L}_1$ that includes the point L_1. Next, we are going to show that the distance between the sensors converges to the constant s. To show this at time kT, we define $d_{\max}(kT) := \max_{1=1,2,\ldots,n} d(p_i(kT), \mathscr{L}_1)$ as the largest distance from the line $\mathscr{L}_1$ to $p_i(kT)$, $i = 1, 2, \ldots, n$. For a given $\delta > 0$, (9.50) implies that there exists $\mathscr{J} \geq 0$ such that $d_{\max}(kT) < \delta$ for all $k \geq \mathscr{J}$. Let $q_i(kT)$ be the projection of the position of sensor i on the line $\mathscr{L}_1$ at time kT, namely $q_i(kT) = l(\bar{\phi} - \pi/2)^T p_i(kT)$. Thus, the control (9.31) along $\mathscr{L}_1$ guarantees that there exists a permutation $\{z_2^{(1)}, z_3^{(1)}, \ldots, z_n^{(1)}\}$ of the set $\{2, 3, \ldots, n\}$ such that

$$q_{z_1^{(1)}} < q_{z_2^{(1)}}(kT) < \ldots < q_{z_n^{(1)}}(kT) \tag{9.51}$$

for all $k \geq \mathscr{J}$, where $z_1^{(1)} = 1$ and $q_{z_1^{(1)}} := l(\bar{\phi} - \pi/2)^T p_1$. The condition (9.51) holds for almost all initial sensor positions. The rules for $\mathscr{D}_i(kT)$, that govern $q_{z_1^{(1)}}(kT)$, $q_{z_3^{(1)}}(kT)$, $\ldots$, $q_{z_n^{(1)}}(kT)$, can be written as a linear dynamic system

$$q^{(1)}((k+1)T) = A^{(1)} q^{(1)}(kT) + b^{(1)}(kT) \quad \text{for } k \geq \mathscr{J} \tag{9.52}$$

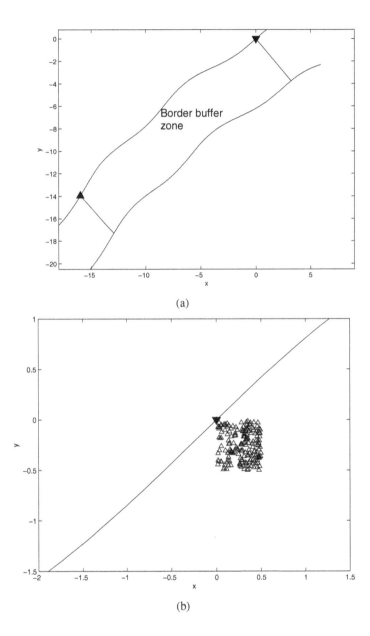

(a)

(b)

Figure 9.14   (a) A country's border buffer zone ($\blacktriangledown$-L_1, $\blacktriangle$-L_2); (b) Initial sensor positions for triangular blanket coverage of the border buffer zone ($\triangle$-initial position, $\blacktriangledown$-L_1).

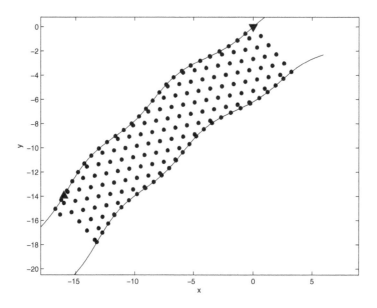

Figure 9.15   Final sensor positions for triangular blanket coverage ($n = 180$, $r_s = 1/\sqrt{3}$) of the border buffer zone ($\bullet$-final position, $\blacktriangledown$-L_1, $\blacktriangle$-L_2).

where

$$q^{(1)}(kT) := \begin{bmatrix} q_{z_2^{(1)}}(kT) & q_{z_3^{(1)}}(kT) & \cdots & q_{z_n^{(1)}}(kT) \end{bmatrix}^T,$$

$$b^{(1)}(kT) := \begin{bmatrix} (q_{z_1^{(1)}} + r_{z_2^{(1)}}(kT))/2 & 0 & \cdots & 0 & s/2 \end{bmatrix}^T$$

and $A^{(1)}$ is a square matrix with elements $A_{i,j}^{(1)}$, $1 \le i, j \le n-1$ such that $A_{i,i+1}^{(1)} = A_{i+1,i}^{(1)} = 1/2$ for $2 \le i \le n-2$; $A_{1,1}^{(1)} = A_{2,1}^{(1)} = A_{n-1,n-1}^{(1)} = 1/2$; and $A_{i,j}^{(1)} = 0$, for all other i, j. For the first layer of sensor array, we have $r_{z_2^{(1)}}(\cdot) \equiv \bar{r}_{z_2^{(1)}} = s$. By using the result [81, p.514] and the fact that $A_{1,1}^{(1)} = A_{2,1}^{(1)} = A_{n-1,n-1}^{(1)} = 1/2$, it can be shown that

$$\lim_{k \to \infty} q_{z_i^{(1)}}(kT) = q_{z_1^{(1)}} + \bar{r}_{z_2^{(1)}} + (i-2)s, = q_{z_1^{(1)}} + (i-1)s$$

for $i = 2, 3, \ldots, n$. Since $\lim_{k \to \infty} d(p_i(kT), \mathscr{L}_1) = 0$, we therefore have

$$\lim_{k \to \infty} p_{z_i^{(1)}}(kT) = L_1 + s(i-1)l(\bar{\phi} - \pi/2) \quad \text{for } i = 2, 3, \ldots, n. \tag{9.53}$$

So far we have showed that if the line $\mathscr{W}_2$ is absent, equation (9.53) does hold. However, when the line $\mathscr{W}_2$ is present, the control law determined by (9.31) and (9.36) drives the sensors that are supposed to lie beyond $\mathscr{W}_2$ to another layer, and these excess sensors form another layer of the sensor array. The distance between this second layer and the first layer is $3r_s/2$. Also, the algorithm places a sensor at h_{1,n_1} on $\mathcal{W}_2$ for the first layer. By the definition of n_1 (i.e., the number of sensors on $\mathcal{L}_1$), there exists a set

$$Z^{(1)} := \{z_1^{(1)}, z_2^{(1)}, \ldots, z_{n_1}^{(1)}\} \subset \{1, 2, \ldots, n\}$$

and $z_1^{(1)} = 1$ such that $\lim_{k \to \infty} \|p_{z_j^{(1)}}(kT) - h_{1,j}\| = 0$ for $j = 1, 2, \ldots, n_1$.

As was mentioned above, the number n_2 of the excess $n - n_1$ sensors then form a second layer of the sensor array. To show this, we repeat the above argument for the first layer of sensors, but in the opposite direction that starts from $\mathcal{W}_2$, to show that n_2 of the $n - n_1$ number of sensors converge to $h_{2,1}, h_{2,2}, \ldots, h_{2,n_2}$. Again, suppose that $\mathcal{W}_1$ is absent. By following the procedure used for the first layer, we can show that the control law (9.37) guarantees that there exists a line

$$\mathcal{L}_2 = \{p \in \mathbb{R}^2 : l(\bar{\phi})^T p = l(\bar{\phi})^T b_2\} \tag{9.54}$$

such that

$$\lim_{k \to \infty} d(p_i(kT), \mathcal{L}_2) = 0, \quad i \in \{1, 2, \ldots, n\} \backslash Z^{(1)}, \tag{9.55}$$

since sensor $z_{n_1}^{(2)}$ converges to the point b_2 and is treated as the leader for the second layer of the sensor array. Also, there exists a permutation $\{z_1^{(2)}, z_2^{(2)}, \ldots, z_{n_2}^{(2)}, \ldots, z_{n-n_1}^{(2)}\}$ of the set $\{1, 2, \ldots, n\} \backslash Z^{(1)}$ and an integer $\mathcal{J}_2$ such that

$$q_{z_1^{(2)}}(kT) > q_{z_2^{(2)}}(kT) > \ldots > q_{z_{n-n_1}^{(2)}}(kT) \tag{9.56}$$

for all $k \geq \mathcal{J}_2$. In fact, sensor $z_1^{(2)}$ is connected with sensor $z_{n_1}^{(1)}$ and is attached to $\mathcal{W}_2$, and the location of this sensor is the starting point for forming the second layer of the sensor array. Similar to (9.52), we can write a linear dynamic system

$$q^{(2)}((k+1)T) = A^{(2)}q^{(2)}(kT) + b^{(2)}(kT) \quad \text{for } k \geq \mathcal{J}_2,$$

where

$$q^{(2)}(kT) := \begin{bmatrix} q_{z_2^{(2)}}(kT) & q_{z_3^{(2)}}(kT) & \cdots & q_{z_{n-n_1}^{(2)}}(kT) \end{bmatrix}^T,$$

$$b^{(2)}(kT) := \begin{bmatrix} (q_{z_1^{(2)}}(kT) - r_{z_2^{(2)}}(kT))/2 & 0 & \cdots & 0 & s/2 \end{bmatrix}^T,$$

and $A_{i,i+1}^{(2)} = A_{i+1,i}^{(2)} = 1/2$ for $2 \leq i \leq (n - n_1) - 2$; $A_{1,1}^{(2)} = A_{2,1}^{(2)} = A_{n-1,n-1}^{(2)} = 1/2$; and $A_{i,j}^{(2)} = 0$ for all other i, j. Since the sensors in the first layer of the sensor array converge to $h_{1,1}, h_{1,2}, \ldots, h_{1,n_1}$, it follows that $r_{z_2^{(2)}}(kT)$ converges to a constant $\bar{r}_{z_2^{(2)}}$ as $r_{z_2^{(2)}}(kT)$ depends on the location of sensor $z_{n_1-1}^{(1)}$. This constant basically determines how much the second row of sensors on $\mathcal{L}_2$ should shift to the left from $\mathcal{W}_2$ (except sensor $z_1^{(2)}$ that is already attached to $\mathcal{W}_2$) so that the interlacing feature of the triangular pattern is achieved. Also, we have $q_{z_1^{(2)}}(kT) \to l(\bar{\phi} - \pi/2)^T b_2$. Therefore,

$$\lim_{k \to \infty} q_{z_i^{(2)}}(kT) = l(\bar{\phi} - \pi/2)^T b_2 - \bar{r}_{z_2^{(2)}} - (i-2)s$$

for $i = 2, 3, \ldots, n - n_1$. Since $\lim_{k \to \infty} d(p_i(kT), \mathscr{L}_2) = 0$, we have

$$
\begin{aligned}
&\lim_{k \to \infty} p_{z_1^{(2)}}(kT) = b_2, \\
&\lim_{k \to \infty} p_{z_i^{(2)}}(kT) = b_2 - (\bar{r}_{z_2^{(2)}} + s(i-2))l(\bar{\phi} - \pi/2))
\end{aligned}
\tag{9.57}
$$

for $i = 2, 3, \ldots, n - n_1$ if $\mathscr{W}_1$ is absent. On the other hand, if $\mathscr{W}_1$ is present, the $n - (n_1 + n_2)$ excess sensors form another layer of the sensor array. Again, the distance between this third layer and the second layer is $3r_s/2$. The last sensor in the second array, namely sensor $z_{n_2}^2$, is driven in such a way that $p_{z_{n_2}}^2(kT) \to a_2 \in \mathscr{W}_1$. Therefore, for the second layer, there exists a set

$$
Z^{(2)} := \{z_1^{(2)}, z_2^{(2)}, \ldots, z_{n_2}^{(2)}\} \subset \{1, 2, \ldots, n\} \backslash Z^{(1)}
$$

such that $\lim_{k \to \infty} \|p_{z_j^{(2)}}(kT) - h_{2, n_2 - j + 1}\| = 0$ for $j = 1, 2, \ldots, n_2$. By following the above argument for showing the formation of the first layer of sensor array, it is then straightforward to show that there exists a set

$$
Z^{(3)} := \{z_1^{(3)}, z_2^{(3)}, \ldots, z_{n_3}^{(3)}\} \subset \{1, 2, \ldots, n\} \backslash (Z^{(1)} \cup Z^{(2)})
$$

such that $\lim_{k \to \infty} \|p_{z_j^{(3)}}(kT) - h_{3, j}\| = 0$ for $j = 1, 2, \ldots, n_1$. Here, sensor $z_1^{(3)}$ is connected with sensor $z_{n_2}^{(2)}$ that is in the second row of the sensor array and is attached to $\mathscr{W}_1$. Therefore, by induction, there always exists a set $Z^{(i)}$ such that

$$
\begin{aligned}
Z^{(1)} &:= \{z_1^{(1)}, z_2^{(1)}, \ldots, z_{n_1}^{(1)}\} \subset \{1, 2, \ldots, n\} \quad \text{for } i = 1, \\
Z^{(i)} &:= \{z_1^{(i)}, z_2^{(i)}, \ldots, z_{n_i}^{(i)}\} \\
&\subset \{1, 2, \ldots, n\} \backslash (Z^{(1)} \cup Z^{(2)} \cup \cdots \cup Z^{(i-1)})
\end{aligned}
$$

for $i = 2, 3, \ldots, K$, such that if i is odd, then

$$
\lim_{k \to \infty} \|p_{z_j^{(i)}}(kT) - h_{i, j}\| = 0, \quad j = 1, 2, \ldots, n_i; \quad \text{or}
$$

$$
\lim_{k \to \infty} \|p_{z_j^{(i)}}(kT) - h_{i, n_i - j + 1}\| = 0, \quad j = 1, 2, \ldots, n_i
$$

if i is even. Hence, condition (9.2) is satisfied, and this completes the proof of Theorem 9.2.

CHAPTER 10

DISTRIBUTED NAVIGATION FOR SWARMING WITH A GIVEN GEOMETRIC PATTERN

10.1 Introduction

Thanks to rapid advances in communication, sensor, and computing technologies and in microelectronics, it becomes more and more popular to deploy multiple mobile robots to cooperatively carry out operations like object transportation [128], localization and mapping [27], and monitoring. Such operations often require that the multi-robot team gets into and subsequently maintains a desired geometric configuration, while all robots move in a common direction with the same speed.

To address this issue, this chapter considers a problem of decentralized self-deployment of a robotic mobile sensor network in a swarm with a prespecified geometric pattern. Unlike Chapter 7, where the objective was to achieve a static deployment of the sensors into a steady geometric structure, now the sensors should move at a prespecified common speed while maintaining a given configuration.

In this chapter, as everywhere in this book, the proposed solution is based on a consensus or agreement scheme and employs local coordination of the team members via the nearest neighbors rules. One of the advantages of this consensus-based approach as compared to, for example, the traditional leader follower formation control (see, e.g., [33, 123] and the literature therein) is that any leader follower scheme

Decentralized Coverage Control Problems for Mobile Robotic Sensor and Actuator Networks. **157**
By Andrey V. Savkin, Teddy M. Cheng, Zhiyu Xi, Faizan Javed, Alexey S. Matveev, and Hung Nguyen. Copyright © 2015 by the Institute of Electrical and Electronics Engineers, Inc.

not only requires but also heavily depends on a single team member appointed as a leader, which may be unwelcome or even unacceptable, especially in adverse circumstances. For example, due to the lack of an explicit feedback to the leader from the followers, the leader moves independently and may walk away and leave its followers behind. On the positive side, the attractiveness of the leader follower scheme is that formation coordination problems can be easily reformulated in the form of well-studied standard regulation or tracking control problems.

Instead of selecting a "physical" leader from the group, a virtual leader can also be employed in formation control. For example, the concept of a virtual leader underlies the decentralized control law for the formation control or stabilization of a group of unicycles offered in [105]. At the same time, the work [105] assumes that the inter-vehicle communication graph is fixed, i.e., the set of neighbors of each vehicle is time-invariant, whereas this graph varies over time in typical consensus schemes. Another approach that is widely adopted for formation control is based on the design and use of artificial potential functions that represent and realize the inter-vehicle interactions and the interactions with the environment; see, e.g., [40,61]. An advantage of this approach is that it naturally leads to a distributed control law and requires only small information exchange between the team members. However, most potential-based algorithms suffer from the fact that it is difficult to guarantee convergence to the desired formation pattern.

The objective of this chapter is to develop a decentralized or distributed navigation strategy based on the consensus approach. Under this strategy, a network of mobile robotic sensors should move in a prespecified geometric pattern from any initial positions and should subsequently stay in the respective formation, while moving with a given speed. For conciseness and definiteness, we focus on a rectangular pattern. At the same time, the proposed control strategy can be modified to achieve other geometric patterns, e.g., triangular or diamond patterns. According to this strategy, no leader is assigned a priori, and autonomous sensors coordinate with each other relying on some global consensus in order to achieve and maintain a rectangular pattern.

Potential applications of the proposed formation control strategy include, but are not limited to, sweep and encircling coverage in missions like minesweeping [12], boarder patrolling [59], environmental monitoring of disposal sites on the deep ocean floor [54], and sea floor surveying for hydrocarbon exploration [9].

The main results of this chapter are originally published in [22].[1]

The rest of the chapter is organized as follows. In Section 10.2, we formulate the problem of decentralized formation control of a network of mobile sensors. An algorithm to address the posed distributed navigation problem is developed in Section 10.3. Section 10.4 presents simulation results to illustrate the proposed algorithm. Theoretical developments of this chapter are given in Section 10.5.

---

[1]The figures of this chapter are reprinted from Cheng, T. M. and Savkin, A. V.: Decentralized control of multi-agent systems for swarming with a given geometric pattern. Computers and Mathematics with Applications. 61(4), 731–744 (2011). Copyright ©2011 Elsevier, with permission from Elsevier.

10.2   Navigation for Swarming Problem

We consider a robotic network in a plane. The network consists of $n \geq 3$ mobile sensors, labelled 1 through n. The objective is to equip every sensor with an autonomous motion control law so that from any initial deployment, the entire network moves in a rectangular lattice pattern and rigidly maintains it afterwards, moving with a prespecified speed. Any sensor can employ only local information.

To specify the problem, we introduce the following notations:

$v_i(\cdot)$	— the speed of sensor i;
$\theta_i(\cdot)$	— the heading of sensor i;
$x_i(\cdot), y_i(\cdot)$	— the abscissa and ordinate, respectively, of sensor i;
$p_i(\cdot) = \begin{bmatrix} x_i(\cdot) \\ y_i(\cdot) \end{bmatrix}$	— the position of sensor i;
r	— the communication range of the sensors;
$l(\alpha) = \begin{bmatrix} \cos \alpha \\ \sin \alpha \end{bmatrix}$	— the unit vector that subtends the angle α with the x-axis;
$T > 0$	— the sampling period;
$s_1, s_2 > 0$	— the dimensions of the rectangular lattice cell;
$\bar{\phi}$	— the orientation angle of the lattice;
$v_0 > 0$	— the speed of the lattice;
$\bar{\psi}$	— the bearing of the lattice velocity;
$\|\cdot\|$	— the Euclidean norm in the plane.

Any mobile sensor is considered as a self-propelled particle with a holonomic drive mechanism, so that the speed v_i and heading θ_i are the control inputs of sensor i. The discrete-time kinematic equation of sensor i is given by:

$$p_i[(k+1)T] = p_i(kT) + v_i(kT)l[\theta_i(kT)]T, \qquad k = 0, 1, 2, \ldots \qquad (10.1)$$

For any sensor γ at any time $t = kT$, we introduce a uniform infinite rectangular lattice with dimensions s_1 and s_2 that is aligned with the angle $\bar{\phi}$ and hosts the current position of the sensor. The vertices of this lattice can be enumerated by pairs (i, j) of integers and are given by

$$h_{i,j}^{\gamma}(kT) = p_{\gamma}(kT) + s_2(i-1)l(\bar{\phi}) + s_1(j-1)l(\bar{\phi} - \pi/2). \qquad (10.2)$$

After this we make this lattice finite by dropping all vertices except for finitely many ones. This operation is regulated by an integer $\bar{K} \in \{1, \ldots, n\}$, which determines the integer quotient p and remainder q of n when divided by $\bar{K}$ (i.e., $n = p\bar{K} + q, 0 \leq q < \bar{K}$). Specifically we retain only n vertices that spread out from the sensor at hand and are organized in p rows if $q = 0$, each with $\bar{K}$ vertices, and in $(p+1)$ rows if $q \neq 0$, where the additional row is incomplete and consists of q vertices; see Fig. 10.1. The retained vertices $h_{i,j}^{\gamma}(kT)$ correspond to the following indices (i, j)

$$i = 1, 2, \ldots, p, \ j = 1, \ldots \bar{K}$$

$$\text{and if } q > 0, \text{ also } i = p+1, j = \begin{cases} 1, \ldots, q & \text{if } i \text{ is odd} \\ \bar{K} - q + 1, \ldots, \bar{K} & \text{if } i \text{ is even} \end{cases}. \qquad (10.3)$$

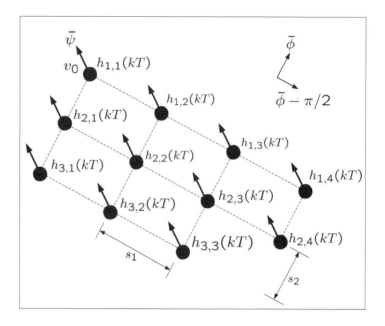

Figure 10.1   A network of mobile sensors moving in the rectangular formation with speed v_0 and heading $\bar{\psi}$ ($n = 11$, $\bar{K} = 4$).

The retained part is also called *lattice* for the sake of brevity and contains the position of the sensor at hand $h_{11}^{\gamma}(kT) = p_{\gamma}(kT)$. It also has a room to accommodate the entire network in its vertices. Any deployment of the remaining $(n-1)$ sensors into the "vacant" vertices is viewed as achievement of the desired geometric pattern.

In the problem to be considered, the integer $\gamma \in \{1, 2, \ldots, n\}$ is not prespecified, and any sensor may eventually take up the $h_{1,1}(kT)$ position. Similarly, the orientation angle $\bar{\phi}$ of the lattice and its distribution over rows and columns, identified with the integer $\bar{K}$, are not given a priori.

The objective is to drive all robots to vertices of such a rectangular lattice and to make them move in the same direction $\bar{\psi}$ at the common speed $v_0 > 0$. Here $\bar{\psi}$ is not given a priory, whereas the desired speed v_0 is prespecified.

Definition 10.1 *A control strategy is called a rectangular formation control if for almost all initial sensor positions, there exists a sensor $\gamma \in \{1, \ldots, n\}$, integer $\bar{K} \in \{1, \ldots, n\}$, and scalars $\bar{\phi}, \bar{\psi} \in [0, \pi)$; and for each pair of indices (i, j) satisfying (10.3), there exists an index $z_{i,j} \in \{1, \ldots, n\}$ such that the following relations hold for any such pair (i, j) and any sensor $s \in \{1, \ldots, n\}$:*

$$\lim_{k \to \infty} \| p_{z_{i,j}}(kT) - h_{i,j}(kT) \| = 0, \tag{10.4}$$

$$\lim_{k \to \infty} \| v_s(kT) - v_0 \| = 0, \qquad \lim_{k \to \infty} \| \theta_s(kT) - \bar{\psi} \| = 0. \tag{10.5}$$

Here "for almost all" means "for all except for a set of zero Lebesgue maesure."

Such a rectangular formation control strategy should operate in a decentralized fashion so that every mobile sensor is driven autonomously on the basis of only local information. Specifically at time $t = kT$, sensor i has access to its own coordinates and also can communicate with its *neighbors*. They are the companion sensors that are currently within the communication range $r > 0$, i.e., in the disk

$$D_{i,r}(kT) := \{p \in \mathbb{R}^2 : \|p - p_i(kT)\| \leq r\}.$$

Via this communication, sensor i may acquire data from its neighbors, but no further information is available on-line.

The relation of neighborship is not steady but may vary over time. To visualize it at time $t = kT$, we introduce an undirected graph $G(kT)$. Its vertices are associated with the sensors, with each vertex being identified by the respective index $i \in \{1,\ldots,n\}$ and called "sensor" for brevity. This graph links vertices $i \neq j$ with and undirected edge if and only if sensors i and j are neighbors at time $t = kT$.

In the study of the posed problem, we adopt the following assumption about the relation of neighborship.

Assumption 10.1 *The graph $G(kT)$ is connected for all $k \geq 0$.*

This assumption is identical to the connectivity assumption adopted in Chapter 9 (i.e., to Assumption 9.1 on page 135), but is stronger than the standard Main Connectivity Assumption from Chapter 2 (i.e., Assumption 2.4 on page 11).

10.3 Distributed Navigation Algorithm

In this section, a decentralized rectangular formation control strategy is presented. This strategy consists of two stages. During the first of them, the sensors align themselves into a straight line formation and each of them acquires identity (ID) within this formation. At this stage, the sensors also come to a consensus about the numbers of rows and columns in the finite lattice, as well as about its orientation and velocity bearing, which are given by $\bar{K}, \bar{\phi}$, and $\bar{\psi}$, respectively. Only after this, they start rearrangement into a rectangular lattice formation with these coordinated parameters, which is performed at the second stage of the algorithm.

10.3.1 First Stage

In this stage, the objective is to steer the sensors so that they ultimately move in a line equispaced formation, as is shown in Fig. 10.2(b). Furthermore, the sensors should come to a consensus about the rectangular lattice to be formed at the second stage. In doing so, every sensor should be driven on the basis of only local information, which includes data about itself and messages communicated from the neighbors.

To reach an agreement with the teammates about the lattice, every sensor i memorizes and, at times $t = kT$, updates the following variables:

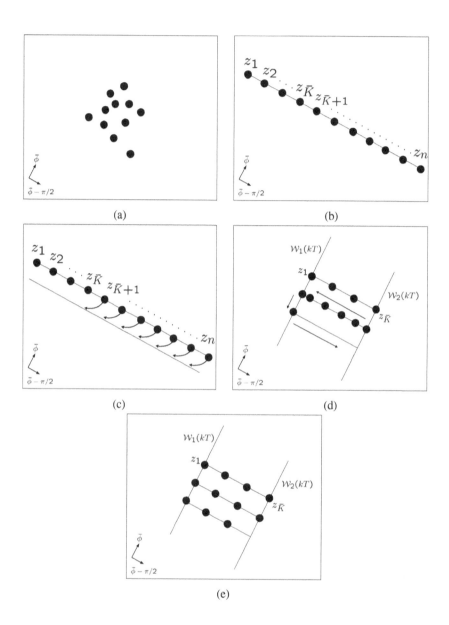

Figure 10.2   Sensors' relative positions: (a) Initial deployment; (b) After stage 1; (c) At the beginning of stage 2; (d) During stage 2; (e) The final rectangular pattern ($n = 11$, $\bar{K} = 4$).

$K_i(kT)$ — the current opinion of sensor i about the value of $\bar{K}$;
$\phi_i(kT)$ — the current opinion of sensor i about the lattice orientation;
$\psi_i(kT)$ — the current opinion of sensor i about the direction of the lattice motion;
$r_i(kT)$ — the current opinion of sensor i about its row in the lattice.

At every time $t = kT$, sensor i broadcasts its coordinates and the current values of these variables and, in return, receives the respective values from its neighbors. Initially, $K_i(0)$ and $\phi_i(0)$, $\psi_i(0)$ are arbitrarily chosen from $\{1,\ldots,n\}$ and $[0, \pi)$, respectively; whereas $r_i(0)$ is set to 1. The choices of different sensors are not coordinated with each other.

During the first stage, the variable $\eta(\cdot)$ is not altered, i.e., all sensors affiliate themselves to the first row. The other variables are updated in accordance with the standard "nearest neighbors rule." Specifically, let $\mathscr{N}_i(kT)$ denote the set of all neighbors $j \neq i$ of sensor i at time $t = kT$, and let $|\mathscr{N}_i(kT)|$ be the number of elements in this set (0 if the set is empty). As usual, the symbol $\lfloor a \rfloor$ stands for the integer floor of a real $a \in \mathbb{R}$, i.e., the greatest integer that does not exceed a. The above updates are carried out as follows:

$$K_i[(k+1)T] := \left\lfloor \frac{1}{1 + |\mathscr{N}_i(kT)|} \left(K_i(kT) + \sum_{j \in \mathscr{N}_i(kT)} K_j(kT) \right) \right\rfloor, \qquad (10.6)$$

$$\psi_i[(k+1)T] := \frac{1}{1 + |\mathscr{N}_i(kT)|} \left(\psi_i(kT) + \sum_{j \in \mathscr{N}_i(kT)} \psi_j(kT) \right), \qquad (10.7)$$

$$\phi_i[(k+1)T] := \frac{1}{1 + |\mathscr{N}_i(kT)|} \left(\phi_i(kT) + \sum_{j \in \mathscr{N}_i(kT)} \phi_j(kT) \right). \qquad (10.8)$$

(The sum over the empty set is defined to be zero, which remark addresses the case where $\mathscr{N}_i(kT) = \emptyset$.) It follows that whereas $K_i(kT)$ takes values only from the finite set $\{1,\ldots,n\}$, the variables $\psi_i(kT)$ and $\phi_i(kT)$ may take any value in the interval $[0, \pi)$ rather than from a discrete set.

To compute the current control $v_i(kT), \theta_i(kT)$, sensor i first generates the set of neighbors that are affiliated to the same row as this sensor, i.e., the set $\mathscr{S}_i(kT) = \{j \in \mathscr{N}_i(kT) : r_j(kT) = r_i(kT)\}$. Then sensor i "projects" the positions of these "row-mates" and its own position onto the direction that is, in its opinion, aligned with the orientation of the lattice by computing the following:

$$c_{ij}(kT) = l\,[\phi_i(kT)]^T\, p_j(kT), \qquad j \in \mathscr{S}_i(kT) \cup \{i\}. \qquad (10.9)$$

The obtained projections are averaged similarly to (10.8):

$$\mathscr{M}_i(kT) := \frac{1}{1 + |\mathscr{S}_i(kT)|} \left(c_{ii}(kT) + \sum_{j \in \mathscr{S}_i(kT)} c_{ij}(kT) \right). \qquad (10.10)$$

The control inputs will be chosen so that at time $t = (k+1)T$ sensor i arrives at the following line translated by the vector $v_0 T l[\psi_i(kT)]$, with the idea in mind that this line is intended to presumably host the entire row at $t = (k+1)T$,

$$\mathscr{L}_i(kT) = \{p \in \mathbb{R}^2 : l[\phi_i(kT)]^T p = \mathscr{M}_i(kT)\}. \tag{10.11}$$

It remains to specify the position of the sensor at hand on this line. To this end, the positions of the same sensors as above are projected onto the direction of this line by computing

$$q_j^i(kT) = l[\phi_i(kT) - \pi/2]^T p_j(kT) \quad \text{for} \quad j \in \mathscr{S}_i(kT) \cup \{i\}.$$

From the so obtained projections, sensor i first selects its own one $q_i^i(kT)$ and then tries to find the projections $q_\alpha^i(kT), q_\beta^i(kT) \neq q_i^i(kT)$ that are left- and right-most closest to $q_i^i(kT)$

$$q_\alpha^i(kT) < q_i^i(kT) < q_\beta^i(kT). \tag{10.12}$$

This attempt succeeds if there are projections to both the left and right of $q_i^i(kT)$. Otherwise only one or even none of these two projections exists. The position of the sensor on the line $\mathscr{L}_i(kT)$ is then determined by its projection $\mathscr{Q}_i(kT)$ onto the respective direction, which projection is defined to be

$$\mathscr{Q}_i(kT) = \frac{q_\alpha^i(kT) + q_\beta^i(kT)}{2} \tag{10.13}$$

if both sensors α and β exist; otherwise, to be

$$\mathscr{Q}_i(kT) = \frac{q_\alpha^i(kT) + q_i^i(kT) + s_1}{2} \tag{10.14}$$

if α exists but β does not; otherwise, to be

$$\mathscr{Q}_i(kT) = \frac{q_\beta^i(kT) + q_i^i(kT) - s_1}{2} \tag{10.15}$$

if β exists but α does not. If none of these cases holds, i.e., neither α nor β exists,

$$\mathscr{Q}_i(kT) := q_i^i(kT).$$

By the foregoing, the desired position of vehicle i at the next sampling time $t = (k+1)T$ is uniquely determined. It remains to find the control inputs that ensure arrival at this position. To this end, the sensor first computes

$$\bar{v}_i(kT) = \frac{\mathscr{Q}_i(kT) - q_i^i(kT)}{T} + v_0 \sin[\phi_i(kT) - \psi_i(kT)],$$

$$\hat{v}_i(kT) = \frac{\mathscr{M}_i(kT) - c_{ii}(kT)}{T} + v_0 \cos[\phi_i(kT) - \psi_i(kT)] \tag{10.16}$$

and then finalizes generation of the current controls by putting

$$v_i(kT) = \sqrt{\bar{v}_i(kT)^2 + \hat{v}_i(kT)^2};$$

$$\theta_i(kT) = \begin{cases} \phi_i(kT) + \xi_i(kT) - \pi/2 & \text{if } \hat{v}_i(kT) \geq 0, \\ \phi_i(kT) - \xi_i(kT) - \pi/2 & \text{if } \hat{v}_i(kT) < 0, \end{cases} \tag{10.17}$$

where $\xi_i(kT) := \cos^{-1} \frac{\bar{v}_i(kT)}{v_i(kT)}$. Under these controls, vehicle i arrives at the position:

$$p_i[(k+1)T] = p_i[kT] + [\mathcal{M}_i(kT) - c_{ii}(kT)]\, l[\phi_i(kT)]$$
$$+ \left[\mathcal{Q}_i(kT) - q_i^i(kT)\right] l\left[\phi_i(kT) - \frac{\pi}{2}\right] + T v_0 l[\psi_i(kT)]. \quad (10.18)$$

10.3.2 Second Stage

At the first stage of the algorithm, the network of mobile sensors is moved into a line formation. However, it ultimately should be driven into a rectangular lattice pattern, as is shown in Fig. 10.1. The objective of the second stage of the algorithm is to perform respective rearrangement from the line formation into this pattern.

As was remarked in Section 10.2, the orientation of the lattice, the bearing of its velocity, and the number of rows are not predefined. These parameters become known only when a global consensus is reached upon their values. Under the connectivity Assumption 10.1, the employed nearest neighbor rules (10.6)—(10.8) guarantee reaching such a consensus, whereas the control (10.17) ensures that all sensors will be ultimately aligned and moving in a line formation.

Once the sensors have been aligned and $K_i(kT)$ has reached a consensus value $\bar{K} \in \{1, 2, \ldots, n\}$, the leftmost sensor initiates a counting sequence, and the sensors then count from the leftmost sensor to the rightmost sensor. As a result, each sensor i acquires a unique ID given by the number $N_i \in \{1, 2, \ldots, n\}$ of counts required to arrive at its position from the leftmost sensor. Let γ and δ stand for the leftmost and rightmost sensors, respectively. Then $N_\gamma = 1, N_\delta = n$, and the number N_i runs from 2 to $n - 1$ as the sensors between γ and δ are observed from left to right. As soon as the counting process has been completed and all the sensors have acquired identification, the algorithm goes to the second stage. Thus at the very beginning of this stage, there exists a permutation $\{z_1, z_2, \ldots, z_n\}$ of the set $\{1, 2, \ldots, n\}$ such that sensor $z_1 = \gamma$ occupies the leftmost position on the line, z_2 is right-closest to sensor z_1, similarly, z_3 is right-closest to sensor z_2, and so on. (In fact, the permutation $\{z_1, z_2, \ldots, z_n\}$ is inverse to $\{N_1, N_2, \ldots, N_n\}$.) In the course of the second stage, the sensors with $N_i > \bar{K}$ move down and form one or more lines that are parallel to the initial line of sensors, as is illustrated in Figs. 10.2(c)—(e), so that the inter-line spacing equals s_2.

The second stage is initiated with transmission of the coordinates of sensor $z_{\bar{K}}$ to sensor z_n via $z_{\bar{K}+1}, \ldots, z_{n-1}$. Meanwhile, sensors $z_{\bar{K}+1}, z_{\bar{K}+2}, \ldots, z_{n-1}, z_n$ set their "row" variable $r_i(\cdot)$ to 2, thus affiliating themselves with the second row of the lattice. By using the received data, agent z_n moves to the location that is below sensor $z_{\bar{K}}$ and is separated by a distance of s_2. In doing so, sensor z_n obeys the following rules:

$$\mathcal{Q}_{z_n}(kT) = \begin{cases} \frac{q_{z_n}^{z_n}(kT) + q_{z_{n-1}}^{z_n}(kT)}{2} & \text{if } q_{z_{n-1}}^{z_n}(kT) > q_{z_{\bar{K}}}^{z_n}(kT), \\ \frac{q_{z_n}^{z_n}(kT) + q_{z_{\bar{K}}}^{z_n}(kT)}{2} & \text{otherwise,} \end{cases} \quad (10.19)$$
$$\mathcal{M}_{z_n}(kT) = \frac{c_{z_n z_n}(kT) + c_{z_n, \bar{K}}(kT) - s_2}{2},$$

where $c_{z_n,j}(\cdot)$ is defined in (10.9). As a result, the position of sensor z_n will satisfy

$$\lim_{k \to \infty} \left[p_{z_{\bar{K}}}(kT) - p_{z_n}(kT) \right] = s_2 l(\bar{\phi}). \tag{10.20}$$

Meanwhile, sensors $z_{\bar{K}+1}, z_{\bar{K}+2}, \ldots, z_{n-1}$ follow sensor z_n to go below the initial line of the sensors since the "row" variables $r_{z_{\bar{K}+1}}(\cdot), r_{z_{\bar{K}+2}}(\cdot), \ldots, r_{z_n}(\cdot)$ have been set to 2. Using algorithms that are similar to the ones developed for stage 1, sensors $z_{\bar{K}+1}, z_{\bar{K}+2}, \ldots, z_n$ start to align themselves to form a second straight-line layer of sensors that is parallel to the first one and is separated by a distance of s_2 from it.

If $n = 2\bar{K}$, the second layer contains $\bar{K}$ sensors, and a rectangular formation is formed. However, if $n > 2\bar{K}$, this layer contains excessively many sensors and so a rectangular pattern cannot be achieved without forwarding $n - 2\bar{K}$ excess sensors to further layers. One way to obtain the desired rectangular configuration, as is shown in Fig. 10.1, is that these excess sensors move down in the direction of $-l(\bar{\phi})$ below sensor z_1 instead of moving beyond sensor z_1 and, after this, start to form the third layer like above. So if $n = 3\bar{K}$, three rows, each with $\bar{K}$ sensors, will be formed. If $n > 3\bar{K}$, the $n - 3\bar{K}$ superfluous sensors are displaced in the direction of $-l(\bar{\phi})$ below sensor z_n; and so on. This process will result in $\lfloor n/\bar{K} \rfloor$ rows, with $\bar{K}$ sensors in each, and possibly the last row with $n - \lfloor n/\bar{K} \rfloor \bar{K}$ sensors, as is illustrated in Fig. 10.2(e); where the last row occurs only if $n - \lfloor n/\bar{K} \rfloor \bar{K} > 0$. (We recall that $\lfloor a \rfloor$ is the integer floor of $a \in \mathbb{R}$, i.e., the maximal integer that does not exceed a.)

To achieve the objective of the second stage, the control inputs should be modified for sensors i whose ID's N_i exceed $\bar{K}$. To this end, two imaginary lines $\mathcal{W}_1(kT)$ and $\mathcal{W}_2(kT)$ are introduced after reaching the second stage; they represent two sides of a rectangle orientated in a direction of $l(\bar{\phi})$. Specifically, the lines are defined as follows (see Fig. 10.2(d)):

$$\begin{aligned} \mathcal{W}_1(kT) &:= \{ p \in \mathbb{R}^2 | (p - p_{z_1}(kT))^T l(\bar{\phi} - \pi/2) = 0 \}, \\ \mathcal{W}_2(kT) &:= \{ p \in \mathbb{R}^2 | (p - p_{z_{\bar{K}}}(kT))^T l(\bar{\phi} - \pi/2) = 0 \}. \end{aligned} \tag{10.21}$$

It is easy to see that they are parallel and are separated by a distance of $(\bar{K} - 1)s_1$.

We first consider sensor i from an even row: $r_i(kT)$ is even. Suppose that it has neighbors to both left and right in the same row, i.e., both α and β are well defined in (10.12). If $p_\alpha(kT) \notin \mathcal{W}_1(kT)$ or $p_\beta(kT) \notin \mathcal{W}_2(kT)$, then $\mathcal{Q}_i(kT)$ is defined by (10.13), as before. If β exists but not α, or if both β and α exist but α is on $\mathcal{W}_1(kT)$, then $\mathcal{Q}_i(kT)$ is defined by (10.15). If α exists but not β, then $\mathcal{Q}_i(kT)$ is given by (10.14). If sensor i hits $\mathcal{W}_1(kT)$ and there currently are no other sensors in $\mathcal{S}_i(kT)$ that lie on $\mathcal{W}_1(kT)$, sensor i is placed on $\mathcal{W}_1(kT)$ by putting $\mathcal{Q}_i(kT) := q^i_{\eta,1}(kT) := q^i_{\eta,1}(kT) := l(\phi_i(kT) - \pi/2)^T \eta_{i,1}(kT)$, where $\eta_{i,1}(kT) := \mathcal{L}_i(kT) \cap \mathcal{W}_1$ and the line $\mathcal{L}_i(kT)$ is defined by (10.11). When sensor i arrives along the line $\mathcal{W}_2(kT)$ from the previous row $r_i(kT) - 1$ and sensors α and β do not exist, sensor i is placed on $\mathcal{W}_2(kT)$ by putting $\mathcal{Q}_i(kT) := q^i_{\eta,2}(kT) := l(\phi_i(kT) - \pi/2)^T \eta_{i,2}(kT)$, where $\eta_{i,2}(kT) := \mathcal{L}_i(kT) \cap \mathcal{W}_2$.

In the case of odd $r_i(kT) \neq 1$, the definition of $\mathcal{Q}_i(kT)$ has much similarity with the case of an even row. If both sensors α and β exist in (10.12), and $p_\alpha(kT) \notin$

$\mathscr{W}_1(kT)$ or $p_\beta(kT) \notin \mathscr{W}_2(kT)$, the quantity $\mathscr{Q}_i(kT)$ is defined by (10.13). If α exists but not β, or if both α and β exist but β is on $\mathscr{W}_2$, then $\mathscr{Q}_i(kT)$ is given by (10.14). If conversely β exists but not α, then $\mathscr{Q}_i(kT)$ is defined by (10.15). If sensor i hits $\mathscr{W}_2(kT)$ and there currently are no other sensors in $\mathscr{S}_i(kT)$ that lie on the line $\mathscr{W}_2(kT)$, sensor i is placed on $\mathscr{W}_2(kT)$ by putting $\mathscr{Q}_i(kT) = q^i_{\eta,2}(kT)$. When sensor i arrives along the line $\mathscr{W}_1(kT)$ from the previous layer $r_i(kT) - 1$ and sensors α and β do not exist, sensor i is placed on $\mathscr{W}_1(kT)$ by putting $\mathscr{Q}_i(kT) = q^i_{\eta,1}(kT)$.

When sensor i is placed on $\mathscr{W}_1(kT)$ with odd $r_i(kT) \neq 1$, or on $\mathscr{W}_2(kT)$ with even $r_i(kT)$, the following quantities are computed:

$$
\begin{aligned}
\mathscr{S}_i(kT) &:= \{j \in \mathscr{N}_i(kT) \mid r_j(kT) = r_i(kT) - 1\}, \\
\gamma_i &:= \underset{j \in \mathscr{S}_i(kT)}{\arg\min} |l(\phi_i(kT))^T (p_j(kT) - p_i(kT))|, \\
\bar{\mathscr{M}}_i(kT) &:= \frac{c_{i,\gamma_i}(kT) + c_{i,i}(kT) - s_2}{2}.
\end{aligned}
\tag{10.22}
$$

The last of them $\bar{\mathscr{M}}_i(kT)$ is used to position the two rows of sensors at a distance of s_2 from each other. By using $\mathscr{M}_i(kT)$, $\bar{\mathscr{M}}_i(kT)$, and $c_{ii}(kT)$, we introduce

$$
\hat{\mathscr{M}}_i(kT) = \begin{cases}
c_{ii}(kT) - s_2 & \text{if } \mathscr{Q}_i(kT) < q^i_{\eta,1}(kT) \\
& \text{or } \mathscr{Q}_i(kT) > q^i_{\eta,2}(kT), \\
\bar{\mathscr{M}}_i(kT) & \text{if } p_i(kT) \in \mathscr{W}_1 \cup \mathscr{W}_2, \\
\mathscr{M}_i(kT) & \text{otherwise,}
\end{cases}
\tag{10.23}
$$

and also

$$
\hat{\mathscr{Q}}_i(kT) = \begin{cases}
q^i_i(kT) & \text{if } \mathscr{Q}_i(kT) < q^i_{\eta,1}(kT) \\
& \text{or } \mathscr{Q}_i(kT) > q^i_{\eta,2}(kT), \\
\mathscr{Q}_i(kT) & \text{otherwise.}
\end{cases}
\tag{10.24}
$$

The row number is updated in accordance with the following rule

$$
r_i((k+1)T) = \begin{cases}
r_i(kT) & \text{if } q^i_{\eta,1}(kT) < \mathscr{Q}_i(kT) < q^i_{\eta,2}(kT), \\
r_i(kT) + 1 & \text{otherwise.}
\end{cases}
\tag{10.25}
$$

Finally the intermediate control variables $\bar{v}_i(kT)$ and $\hat{v}_i(kT)$ in (10.16) are modified as follows:

$$
\begin{aligned}
\bar{v}_i(kT) &= \frac{\hat{\mathscr{Q}}_i(kT) - q^i_i(kT)}{T} + v_0 \sin[\phi_i(kT) - \psi_i(kT)], \\
\hat{v}_i(kT) &= \frac{\hat{\mathscr{M}}_i(kT) - c_{ii}(kT)}{T} + v_0 \cos[\phi_i(kT) - \psi_i(kT)].
\end{aligned}
\tag{10.26}
$$

The final control inputs are still given by (10.17).

10.3.3   Behavior of the Proposed Algorithm

In the remainder of the chapter, we provide an evidence in favor of the following.

Statement 10.1 *Let Assumption 10.1 hold. Then the proposed decentralized control strategy is a rectangular formation control in the sense of Definition 10.1.*

10.4   Illustrative Examples and Computer Simulation Results

In this section, we present numerical simulations illustrating the proposed algorithm and its behavior. Figure 10.3 addresses the first simulation experiment, where nineteen ($n = 19$) mobile sensors eventually form a rectangular formation with the consensus value $\bar{K} = 5$ and move in the consensus direction of $\bar{\psi} = 1.433$. Figure 10.3(a) shows that the sensors were successfully aligned during the first stage. Various episodes of the second stage are depicted in Figs. 10.3(b)–10.3(e); they display a good accordance with the ideal behavioral pattern from Fig. 10.2. Whereas only two rows are formed in Fig. 10.3(b), the sensors are subsequently reorganized into three rows and then four rows, as is shown in Figs 10.3(c) and 10.3(d), respectively. Though the last row has been already formed in the episode from Fig. 10.3(d), the sensors in this and the third row are not yet equally spaced and the inter-sensor distances differ from those in the first and second rows. However, this difference fades away and becomes nearly invisible at time $k = 380$, as is shown in Fig. 10.3(e). The simulation experiment shows that from this time on, the sensors maintain and collectively move in the rectangular formation displayed in Fig. 10.3(e). The incomplete last row with only four sensors is due to their total number $n = 19$ and the consensus value $\bar{K} = 5$. Figure 10.4 exhibits the speed $v_i(\cdot)$ and heading $\theta_i(\cdot)$ profiles of all sensors. As it can be seen, the speeds of all the sensors converge to a common value $v_0 = 0.01$, i.e., the sensors ultimately move with the same speed, as is desired. Similarly, the headings also converge to a common value: All the sensors ultimately move in the same direction ($\bar{\psi} = 1.433$), as is needed.

The first simulation dealt with the basic scenario considered in this chapter: The orientation, velocity heading, and dimensions of the rectangular lattice (given by $\bar{\phi}$, $\bar{\psi}$, and $\bar{K}$, respectively) are unknown to the sensors a priori; these parameters become known only after a decentralized consensus on their values is reached among the sensors. To reach consensus, every sensor forms its own opinion about these parameters and at every sampling time, updates this opinion based on the available local information. Meanwhile, the initial opinion of every sensor is not somehow limited; and the opinions of different sensors may be uncorrelated.

At the same time, the developments of this chapter also cover the converse scenario where the sensors should ultimately form a rectangular lattice with a prespecified orientation, velocity heading, speed, and dimensions (i.e., with a prespecified $\bar{\phi}$, $\bar{\psi}$, v_0, and $\bar{K}$, respectively). To achieve this objective, it suffices to initially supply all the sensors with the desired values of these parameters so that any sensor may use them to set the initial values of the respective coordination variables. Since the initial value of any variable is common for all sensors, the update rules (10.6)—(10.8) do

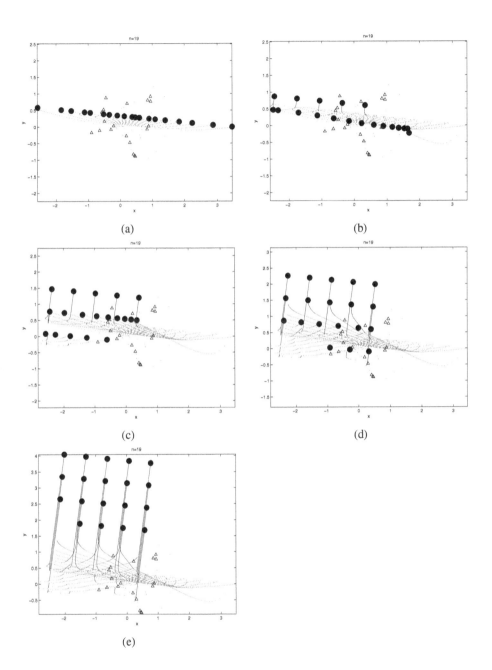

Figure 10.3   Rectangular formation of nineteen ($n = 19$) sensors with consensus value $\bar{K} = 5$: the sensor positions at (a) $k = 30$; (b) $k = 60$; (c) $k = 120$; (d) $k = 240$; and (e) $k = 380$ (the initial positions are shown as $\triangle$).

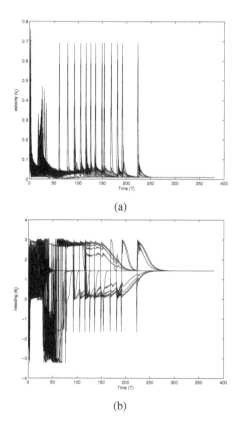

(a)

(b)

Figure 10.4   (a) The speed v_i; (b) The heading θ_i of the sensors.

not change this value afterwards. So consensus will be immediately established and the agreed values the lattice parameters will equal their respective initial values and will be exactly the given and desired ones.

This scenario is addressed in the second simulation. Forty $n = 40$ sensors are required to ultimately move in a direction of $\pi/4 rad$ and form a rectangular lattice with five rows, each with eight sensors; the lattice orientation should be $\pi/4 rad$. To this end, the initial values of $\psi_i(\cdot)$, $\phi_i(\cdot)$ and $K_i(\cdot)$ are set to be $\pi/4 rad$, $\pi/4 rad$, and 8, respectively, for all the sensors. Figures 10.5 and 10.6 show that the control objective is successfully achieved under the proposed control strategy.

Up to now, it was assumed that position observations performed by the sensors are absolutely exact. However, measurement errors are inevitable in practice despite the fact that reliability of observations can be improved via a range of techniques, like the robust Kalman state estimation [86, 87]. Whereas Statement 10.1 addresses the error-free case and perfect convergence of the mobile network to a rectangular lattice moving at the given speed v_0 in an agreed direction $\bar{\psi}$, a highly plausible effect of a bounded nondecaying measurement noise is that convergence holds with only a certain nonzero precision: there exist a finite time $t = Tk_*$ and scalars $\varepsilon_v, \varepsilon_\theta > 0$ such that $|v_i(kT) - v_0| \le \varepsilon_v$ and $|\theta_i(kT) - \bar{\psi}| \le \varepsilon_\theta$ for all $k \ge k_\psi$ and $i = 1, 2, \dots, n$. (Here typically $\varepsilon_v, \varepsilon_\psi \to 0$ as the noises go to zero.) This leaves a room to oscillate or cycle around v_0 and $\bar{\psi}$, respectively.

Exactly such a behavior is confirmed by a simulation study whose results are depicted in Figs 10.7 and 10.8(a). In the respective test, the measurements of positions $x_i(kT)$ and $y_i(kT)$ are corrupted by mutually independent additive noises $\varepsilon_i^x(kT)$ and $\varepsilon_i^y(kT)$, respectively; the both noises are uniformly distributed over $[-0.005, 0.005]$. Figure 10.7 displays visible oscillations of the heading and speed of each sensor around $\bar{\psi} = 1.422$ and the prespecified speed $v_0 = 0.005$, respectively.

To attenuate cycling, a simple heuristic strategy was proposed and validated via computer simulations. First a threshold $v_e > 0$ is chosen that should presumably be (or is experimentally chosen to be) less than ε_v. After the network enters a regime of oscillations around a perfect rectangular lattice, the following rule is additionally applied: if $|v_i(kT) - v_0| \le v_e$, then

$$v_i(kT) := v_0 \quad \text{and} \quad \theta_i(kT) := \psi_i(kT). \tag{10.27}$$

The effect of this rule with $v_e = 0.009$ in the experimental setup from Figs 10.7 and 10.8(a) is demonstrated in Figs 10.8(b) and 10.9. As it can be seen, cycling is really attenuated: The speeds and headings of the sensors are in a sharp contrast to those obtained without the attenuation rule and displayed in Figs 10.7(a) and (b). In other words, the above strategy may make the formation less sensitive to noises.

10.5 Theoretical Analysis of the Algorithm

In the previous section, it was demonstrated via numerous computer experiments that the proposed strategy achieves the control objective. Now we present a theoretical evidence that this success is not occasional but is rigorously predetermined by the

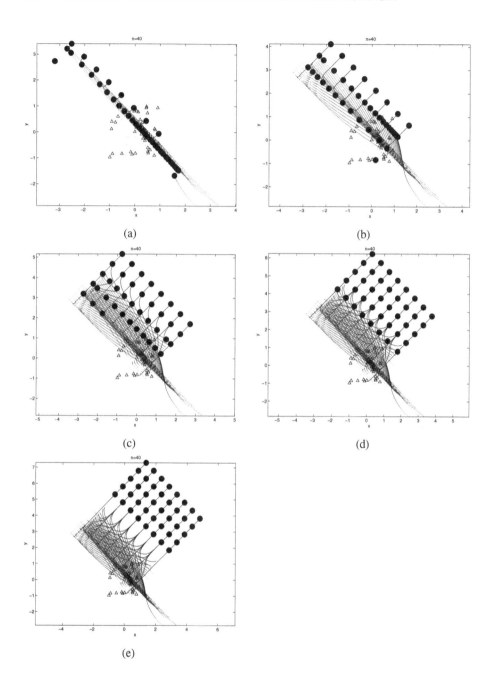

Figure 10.5 Forty sensors move in a rectangular formation with prespecified parameters $\bar{\psi}$, $\bar{\phi}$ and $\bar{K}$: the sensor positions at (a) $k = 100$. (b) $k = 300$. (c) $k = 600$. (d) $k = 900$. (e) $k = 1200$ (the initial positions are shown as $\triangle$).

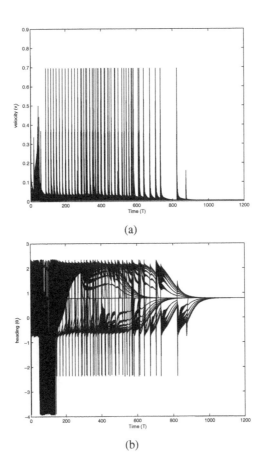

Figure 10.6   (a) The speeds v_i. (b) The headings θ_i. The sensors move in a rectangular pattern with prespecified parameters $\bar{\psi}$, $\bar{\phi}$, and $\bar{K}$.

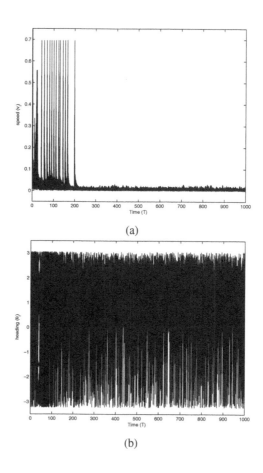

(a)

(b)

Figure 10.7   (a) The speed v_i. (b) The heading θ_i. Noisy position measurements.

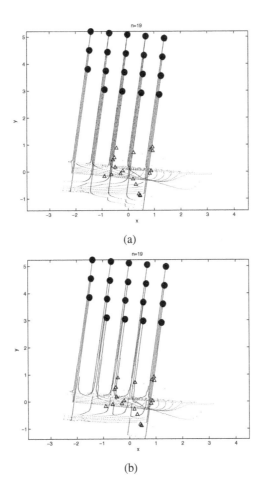

(a)

(b)

Figure 10.8   Nineteen sensors move in a rectangular formation with $\bar{K} = 5$: the snapshots at $k = 1000$ (a) without cycling attenuation strategy and b) with cycling attenuation strategy.

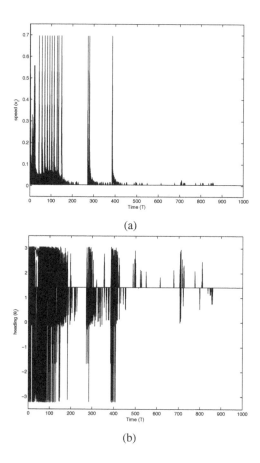

Figure 10.9 (a) The speed v_i. (b) The heading θ_i. The effect of cycling attenuation strategy in the case of noisy position measurements.

logic of the algorithm. The core of the resultant behavior is constituted by alignment into an equispaced formation along a straight line. Such an alignment of the entire team is the goal of the first stage of the algorithm, and alignment within a particular row plays a key role in the second stage, whereas nearly identical rules are applied to achieve these alignments in the both stages. So it is natural to examine first the outcome of these rules. In doing so, we mainly focus on the first stage for the sake of definiteness.

At the start of this stage, sensor i has initial data $\{p_i(0), \theta_i(0), \psi_i(0), \phi_i(0), K_i(0)\}$; these data are not correlated among various sensors. The connectivity property imposed in Assumption 10.1 and the update rules (10.6)—(10.8) imply reaching a consensus: There exist $\bar{\psi}, \bar{\phi} \in [0, \pi)$ and $\bar{K} \in \{1, 2, \ldots, n\}$ such that as $k \to \infty$,

$$\psi_i(kT) \to \bar{\psi}, \quad \phi_i(kT) \to \bar{\phi}, \quad K_i(kT) \to \bar{K} \tag{10.28}$$

for $i = 1, 2, \ldots, n$; see, e.g., [53, 92, 96]. Moreover, this convergence is as fast as exponential [10].

Using $\bar{\psi}$, we define the system

$$\hat{p}((k+1)T) = \hat{p}(kT) + v_0 T l(\bar{\psi}) \tag{10.29}$$

with $\hat{p}(0) = 0$, and for any $i = 1, 2, \ldots, n$ and $k = 0, 1, 2, \ldots$, write $p_i(kT)$ as follows:

$$p_i(kT) = \bar{p}_i(kT) + \hat{p}(kT). \tag{10.30}$$

By invoking (10.18) and the first two relations from (10.28), it is easy to see that the state $\bar{p}_i(kT)$ is governed by

$$\bar{p}_i[(k+1)T] = \bar{p}_i[kT] + [\mathscr{M}_i(kT) - c_{ii}(kT)] l(\bar{\phi})$$
$$+ \left[\mathscr{Q}_i(kT) - q_i^i(kT)\right] l\left[\bar{\phi} - \frac{\pi}{2}\right] + w_i(kT). \tag{10.31}$$

From now on, the symbol $w(kT)$ (with possible indices) is used to denote variables that exponentially converge to zero as $k \to \infty$.

Next, we consider the system (10.31) and suppose that the algorithm does not switch to the second stage. We are going to show that the sensors ultimately come in a formation along a straight line with agreed orientation: There exist a scalar $\bar{c}_1$ and a line

$$\bar{\mathscr{L}}_1 = \{p \in \mathbb{R}^2 : p^T l(\bar{\phi}) = \bar{c}_1\} \tag{10.32}$$

such that

$$\lim_{k \to \infty} d\left[\bar{p}_i(kT), \bar{\mathscr{L}}_1\right] = 0 \quad i = 1, 2, \ldots, n. \tag{10.33}$$

Here $d[p, \mathscr{L}]$ is the distance between the point p and the line $\mathscr{L}$.

To this end, we put, in the vein of (10.9),

$$\bar{c}_j(kT) := l(\bar{\phi})^T \bar{p}_j(kT), \quad j \in \mathscr{S}_i(kT) \cup \{i\};$$
$$\hat{c}_i(kT) := l[\phi_i(kT)]^T \hat{p}(kT). \tag{10.34}$$

Based on (10.9), (10.30), and the second relation from (10.28), we conclude that

$$c_{ij}(kT) = l[\phi_i(kT)]^T [\bar{p}_j(kT) + \hat{p}(kT)] = \bar{c}_j(kT) + \hat{c}_i(kT) + w_{ij}(kT).$$

Hence by invoking (10.10), we see that

$$\mathcal{M}_i(kT) - c_{ii}(kT) = \frac{1}{1 + |\mathcal{S}_i(kT)|} \left(\sum_{j \in \mathcal{S}_i(kT)} [c_{ij}(kT) - c_{ii}(kT)] \right)$$

$$= \frac{1}{1 + |\mathcal{S}_i(kT)|} \left(\sum_{j \in \mathcal{S}_i(kT)} [\bar{c}_j(kT) + \hat{c}_i(kT) + w_{ij}(kT) - \bar{c}_i(kT) - \hat{c}_i(kT) - w_{ii}(kT)] \right)$$

$$= \frac{1}{1 + |\mathcal{S}_i(kT)|} \left(\sum_{j \in \mathcal{S}_i(kT)} [\bar{c}_j(kT) - \bar{c}_i(kT)] \right) + \bar{w}_i(kT).$$

Hence

$$\bar{c}_i[(k+1)T] = l(\bar{\phi})^T \bar{p}_j[(k+1)T] \overset{(10.31)}{=} \bar{c}_i[kT] + \mathcal{M}_i(kT) - c_{ii}(kT) + l(\bar{\phi})^T w_i(kT)$$

$$= \frac{1}{1 + |\mathcal{S}_i(kT)|} \left(\bar{c}_i(kT) + \sum_{j \in \mathcal{S}_i(kT)} \bar{c}_j(kT) \right) + \hat{w}_i(kT).$$

Thus we see that the variables $\bar{c}_1(kT), \ldots, \bar{c}_n(kT)$ solve a linear time-varying nonhomogeneous equation. Given $k_* = 0, 1, \ldots,$ for $k \geq k_*$, this solution can be represented as the sum $\bar{c}_i(kT) = c_i^0(kT) + c_i^-(kT)$ of the solution $c_1^0(kT), \ldots, c_n^0(kT)$ of the respective homogeneous equation

$$c_i^0[(k+1)T] = \frac{1}{1 + |\mathcal{S}_i(kT)|} \left(c_i^0(kT) + \sum_{j \in \mathcal{S}_i(kT)} c_j^0(kT) \right), \tag{10.35}$$

which solution inherits the initial data from $\bar{c}_1(kT), \ldots, \bar{c}_n(kT)$ at $k = k_*$, and the solution $c_1^-(kT), \ldots, c_n^-(kT)$ of the non-homogeneous equation at hand starting at $k = k_*$ with the zero data:

$$c_i^-[(k+1)T] = \frac{1}{1 + |\mathcal{S}_i(kT)|} \left(c_i^-(kT) + \sum_{j \in \mathcal{S}_i(kT)} c_j^-(kT) \right) + \hat{w}_i(kT) \quad k \geq k_*,$$

$$c_1^-(k_*T) = \cdots = c_n^-(k_*T) = 0.$$

Since (10.35) represents the standard "nearest neghbours rule," Assumption 10.1 guarantees achievement of a consensus (see, e.g., [53,92,96]): There exists $c_\infty^0(k_*) \in \mathbb{R}$ such that

$$\lim_{k \to \infty} c_i^0(kT) = c_\infty^0(k_*) \qquad i = 1, \ldots, n.$$

Meanwhile, by putting $\hat{w}(kT) := \max_{i=1,\ldots,n} |\hat{w}_i(kT)|$, it is easy to see (by induction on $k = k_*, k_* + 1, \ldots$) that

$$|c_i^-(kT)| \leq \sum_{l=k_*}^{k-1} \hat{w}(lT) \leq c_\infty^-(k_*) := \sum_{l=k_*}^{\infty} \hat{w}(lT) \quad i = 1, \ldots, n.$$

By bringing the pieces together, we see that

$$\overline{\lim_{k\to\infty}} \bar{c}_i(kT) \le c_\infty^0(k_*) + c_\infty^-(k_*) < \infty,$$

$$\underline{\lim_{k\to\infty}} \bar{c}_i(kT) \ge c_\infty^0(k_*) - c_\infty^-(k_*) > -\infty \tag{10.36}$$

and so

$$\overline{\lim_{k\to\infty}} \bar{c}_i(kT) - \underline{\lim_{k\to\infty}} \bar{c}_i(kT) \le 2c_\infty^-(k_*) \qquad \forall k_* = 0,1,\dots.$$

Now we recall that the symbol $w(kT)$ (with possible indices) is used to denote variables that exponentially converge to zero. It follows that $c_\infty^-(k_*) \to 0$ as $k_* \to \infty$ and so $\overline{\lim}_{k\to\infty} \bar{c}_i(kT) = \underline{\lim}_{k\to\infty} \bar{c}_i(kT)$, i.e., there exists $\lim_{k\to\infty} \bar{c}_i(kT) =: c_i^\infty$. By invoking (10.36) once more, we conclude that $|c_i^\infty - c_j^\infty| \le 2c_\infty^-(k_*)$ for all $k_* = 0,1,\dots$, and letting $k_* \to \infty$ assures that $c_i^\infty = c_j^\infty$ for all $i,j = 1,\dots,n$. Let $\bar{c}$ denote this common value of c_i^∞'s. By the foregoing and (10.34), $l(\bar{\phi})^T \bar{p}_j(kT) \to \bar{c}$ as $k \to \infty$. This means that (10.33) does hold, i.e., the sensors ultimately come in a formation along a straight line.

However the above property does not guarantee equal and required spacing between the sensors, i.e., that there exists a permutation $\{z_1, z_2, \dots, z_n\}$ of the set $\{1, 2, \dots, n\}$ such that

$$\lim_{k\to\infty} \|\bar{p}_{z_i}(kT) - \bar{p}_{z_1}(kT)\| = (i-1)s_1 \qquad i = 1,\dots,n. \tag{10.37}$$

To show (10.37), we start our analysis at time $t = k_*T$ such that for $k \ge k_*$ all the states $\bar{p}_i(kT), i = 1,\dots,n$, are already sufficiently close to the line $\mathcal{L}_1$, and assume a typical for such a concluding phase situation where for $k \ge k_*$, these states are pairwise disjoint after projection onto the direction of $\mathcal{L}_1$, i.e., the quantity

$$\bar{q}_i(kT) := [\sin(\bar{\phi}) \quad -\cos(\bar{\phi})]^T \times \bar{p}_i(kT) \tag{10.38}$$

does not assume a common value for different i's. Then these quantities can be put in the ascending order via proper re-enumeration: There exists a permutation $\{z_1^{(k)}, z_2^{(k)}, \dots, z_n^{(k)}\}$ of the set $\{1, 2, \dots, n\}$ such that

$$\bar{q}_{z_1^{(k)}}(kT) < \bar{q}_{z_2^{(k)}}(kT) < \dots < \bar{q}_{z_n^{(k)}}(kT). \tag{10.39}$$

We also assume another typical feature: Neighborship in this sequence implies that the sensors are neighbors, i.e., lie within the communication range of each other, and the above permutation does not evolve over time, i.e., $\{z_1^{(k)}, z_2^{(k)}, \dots, z_n^{(k)}\} = \{z_1, z_2, \dots, z_n\}$ for $k \ge k_*$. Then in terms of the following vectors from $\mathbb{R}^{n-1}$

$$\Delta(kT) := \begin{bmatrix} \bar{q}_{z_2}(kT) - \bar{q}_{z_1}(kT) \\ \bar{q}_{z_3}(kT) - \bar{q}_{z_2}(kT) \\ \vdots \\ \bar{q}_{z_{n-1}}(kT) - \bar{q}_{z_{n-2}}(kT) \\ \bar{q}_{z_n}(kT) - \bar{q}_{z_{n-1}}(kT) \end{bmatrix}, \quad b := \begin{bmatrix} s_1/2 \\ 0 \\ \vdots \\ 0 \\ s_1/2 \end{bmatrix}$$

and with regard to (10.28), the update rules (10.13)—(10.15) can be re-written as the linear recursion:

$$\Delta[(k+1)T] = A\Delta(kT) + b + w_\Delta(kT), \tag{10.40}$$

where $w_\Delta(kT)$ converges to zero as $k \to \infty$ and

$$A = \begin{pmatrix} 0 & 1/2 & 0 & 0 & 0 & 0 & 0 \\ 1/2 & 0 & 1/2 & 0 & 0 & 0 & 0 \\ 0 & 1/2 & 0 & 1/2 & 0 & 0 & 0 \\ 0 & 0 & 1/2 & 0 & 1/2 & 0 & 0 \\ 0 & 0 & 0 & 1/2 & 0 & 1/2 & 0 \\ 0 & 0 & 0 & 0 & 1/2 & 0 & 1/2 \\ 0 & 0 & 0 & 0 & 0 & 1/2 & 0 \end{pmatrix}.$$

By using the result [81, p.514], it can be shown that all eigenvalues of A lie in the open unit disk and so $\lim_{k\to\infty} \Delta(kT) = \Delta_\infty := (I - A)^{-1}b$. Here Δ_∞ is the unique solution of the equation $\Delta_\infty = A\Delta_\infty + b$; it is easy to see by inspection that $\Delta_\infty = (s_1, s_1, \ldots, s_1)^T$. Hence $\bar{q}_{z_i}(kT) - \bar{q}_{z_{i-1}}(kT) \to s_1$ as $k \to \infty$ for all $i = 2, \ldots, n$, which evidently implies (10.37).

Thus we have provided an evidence that if the algorithm does not switch to the second stage, the sensors will get in a line formation and the distance between sensors in this line formation is s_1. On the other hand, when the second stage is activated, the algorithm drives the sensors that have the ID's greater than $\bar{K}$ to form a number of parallel layers of sensors, with the distance between the layers being s_2.

As discussed before, if $n \geq 2\bar{K}$, the $\bar{K}$ of the excess $n - \bar{K}$ sensors will then form a second layer of the sensor array. Repeating the above arguments, but in the opposition direction below sensor $\bar{K}$ and with taking into account that the starting point for forming the second layer of the sensor array is below the sensor that corresponds to the $h_{1\bar{K}}$-position, shows that $\bar{K}$ of the $n - \bar{K}$ sensors converge to $h_{2,1}(kT), h_{2,2}(kT), \ldots, h_{2,\bar{K}}(kT)$ relative to sensor γ. By continuing, if necessary, likewise with respect to the third, fourth, and so on layers, we arrive at the conclusion that the sensors do converge to the desired rectangular lattice pattern.

REFERENCES

1. Acar E. U., Choset H., Zhang Y. and Schervish M.: Path planning for robotic dem-ining: Robust sensor-based coverage of unstructured environments and probabilistic methods. International Journal of Robotics Research. **22(7–8)**, 441–466 (2003)

2. Al-Takrouri S. and Savkin A. V.: A decentralized flow redistribution algorithm for avoiding cascaded failures in complex networks. Physica A. **392(23)**, 6135–6145 (2013)

3. Al-Takrouri S., Savkin A. V. and Agelidis V. G.: A decentralized control algorithm based on the DC power flow model for avoiding cascaded failures in power networks. Proceedings of the 9th Asian Control Conference (ASCC). (Istanbul, Turkey, 2006)

4. Bai X., Kumar S., Xuan D., Yun Z. and Lai T. H.: Deploying wireless sensors to achieve both coverage and connectivity. Proceedings of the International Symposium on Mobile Ad Hoc Networking and Computing (MobiHoc). 131–142 (2006)

5. Blondel V. D., Hendrickx J. M., Olshevsky A., and Tsitsiklis J. N.: Convergence in multiagent coordination, consensus, and flocking. Proceedings of the IEEE Conference on Decision and Control. 2996 – 3000 (2005)

6. Breder C. M.: Equations descriptive of fish schools and other animal aggregations. Ecology. **35(3)**, 361–370 (1954)

7. Bullo F., Cortés J. and Martínez S., *Distributed Control of Robotic Networks*. (Princeton University Press, Applied Mathematics, 2009, 978-0-691-14195-4)

8. Butler Z., Rizzi A. and Hollis R.: Distributed coverage of rectilinear environments. Workshop on the Algorithmic Foundations of Robotics. (Boston, MA, USA, 2001)

9. Børhaug E., Pavlov A. and Pettersen K. Y.: Straight line path following for formations of underactuated underwater vehicles. Proceedings of the 46th IEEE Conference on Decision and Control. 1905–1912 (New Orleans, USA, 2007)

10. Cao M., Spielman D. A. and Morse A. S.: A lower bound on convergence of a distributed network consensus algorithm. Proceedings of 44th IEEE Conference on Decision and Control. (Seville, Spain, 2005)

11. Cao Y. U., Fukunaga A. S. and Kahng A. B.: Cooperative mobile robotics: Antecedents and directions. Autonomous Robots. 4(1), 7–27 (1997)

12. Cassinis R., Bianco G., Cavagnini A. and Ransenigo P.: Strategies for navigation of robot swarms to be used in landmines detection. Third European Workshop on Advanced Mobile Robots. 22(7–8), 211–218 (Zurich, Switzerland, 1999)

13. Chang C-Y., Chang C-T., Chen Y-C. and Chang H-R.: Obstacle-resistant deployment algorithms for wireless sensor networks. IEEE Transactions on Vehicular Technology. 58(6), 2925–2941 (2009)

14. Chao H., Chen Y. and Ren W.: A study of grouping effect on mobile actuator sensor networks for distributed feedback control of diffusion process using central Voronoi tessellations. Proceedings of IEEE International Conference on Mechatronics and Automation. 25–28, 769–774 (2006)

15. Chellappan S., Gu W., Bai X., Xuan D., Ma B. and Zhang K.: Deploying wireless sensor networks under limited mobility constraints. IEEE Transactions on Mobile Computing. 6(10), 1142–1157 (2007)

16. Chen A., Kumar S. and Lai T.: Designing localized algorithms for barrier coverage. Annual International Conference on Mobile Computing and Networking (MOBICOM). 63–74 (Montreal, Canada, 2007)

17. Chen A., Kumar S. and Lai T. H.: Local barrier coverage in wireless sensor networks. IEEE Transactions on Mobile Computing. 9(4), 491–504 (2010)

18. Cheng C. F., Wu T. Y. and Liao H. C.: A density-barrier construction algorithm with minimum total movement in mobile WSNs. Computer Networks. 62(7), 208–220 (2014)

19. Cheng T. M. and Savkin A. V.: A distributed self-deployment algorithm for the coverage of mobile wireless sensor networks. IEEE Communications Letters. 13(11), 877–879 (2009)

20. Cheng T. M. and Savkin A. V.: Decentralized control of mobile sensor networks for triangular blanket coverage. Proceedings of the American Control Conference. (Baltimore, Maryland, 2010)

21. Cheng T. M. and Savkin A. V.: Decentralized control for mobile robotic sensor network self-deployment: Barrier and sweep coverage problems. Robotica. 29, 283–294 (2011)

22. Cheng T. M. and Savkin A. V.: Decentralized control of multi-agent systems for swarming with a given geometric pattern. Computers and Mathematics with Applications. 61(4), 731–744 (2011)

23. Cheng T. M. and Savkin A. V.: Self-deployment of mobile robotic sensor networks for multilevel barrier coverage. Robotica. 30(4), 661–669 (2012)

24. Cheng T. M. and Savkin A. V.: Decentralized control of mobile sensor networks for asymptotically optimal blanket coverage between two boundaries. IEEE Transactions on Industrial Informatics. **9(1)**, 365–376 (2013)

25. Cheng T. M., Savkin A. V. and Javed F.: Decentralized control of a group of mobile robots for deployment in sweep coverage. Robotics and Autonomous Systems. **59(7–8)**, 497–507 (2011)

26. Choset H.: Coverage for robotics: A survey of recent results. Annals of Mathematics and Artificial Intelligence. **31(1–4)**, 113–126 (2001)

27. Choset H. and Nagatani K.: Topological simultaneous localization and mapping (slam): Toward exact localization without explicit localization. IEEE Transactions on Robotics and Automation. **17(2)**, 125–137 (2001)

28. Cochran, J., Krstic, M.: Nonholonomic source seeking with tuning of angular velocity. IEEE Transactions on Automatic Control. **54(4)**, 717–731 (2009)

29. Cortés J., Martinez S., Karatas T. and Bullo F.: Coverage control for mobile sensing networks. IEEE Transactions on Robotics and Automation. **20(2)**, 243–255 (2004)

30. Das A. K., Fierro R., Kumar V., Ostrowski J. P., Spletzer J. and Taylor C. J.: A vision-based formation control framework. IEEE Transactions on Robotics and Automation. **18**, 813–825 (2002)

31. Demetriou M. A.: Guidance of mobile actuator-plus-sensor networks for improved control and estimation of distributed parameter systems. IEEE Transactions on Automatic Control. **55(7)**, 1570–1584 (2010)

32. Dong D., Liao X., Liu K., Liu Y. and Xu W.: Distributed coverage in wireless ad hoc and sensor networks by topological graph approaches. IEEE Transactions on Computers. **61(10)**, 1417–1428 (2012)

33. Egerstedt M. and Hu X.: Formation constrained multi-agent control. IEEE Transactions on Robotics and Automation. **17(6)**, 947–951 (2001)

34. Flierl G., Grunbaum D., Levin S. and Olson D.: From individuals to aggregations: The interplay between behavior and physics. Journal of Theoretical Biology. **196**, 397–454 (1999)

35. Gage D.W.: Command control for many-robot systems. Proceedings of the 19th Annual AUVS Technical Symposium. **4**, 22–24 (Hunstville, Alabama, USA, 1992)

36. Gage D.W.: Many-robot MCM search systems. Proceedings of Autonomous Vehicles in Mine Countermeasures Symposium. 9–55 (Monterey, CA, USA, 1995)

37. Gallais A., Carle J., Simplot-Ryl D. and Stojmenović I.: Localized sensor area coverage with low communication overhead. IEEE Transactions on Mobile Computing. **7(5)**, 661–672 (2008)

38. Garcia, E. and De Santos P. G.: Mobile-robot navigation with complete coverage of unstructured environments. Robotics and Autonomous Systems. **46(4)**, (2004)

39. Gazi, V. and Fidan, B.: Coordination and control of multi-agent dynamic systems: Models and approaches. Swarm Robotics, Lecture Notes in Computer Science. 71–102 (Springer, Berlin, 2007)

40. Gazi V. and Passino K. M.: A class of attractions/repulsion functions for stable swarm aggregations. International Journal of Control. **77(18)**, 1567–1579 (2004)

41. Ge S. S. and Fua C: Complete multi-robot coverage of unknown environments with minimum repeated coverage. Proceedings of the IEEE International Conference on Robotics and Automation (ICRA). 715–720 (2005)

42. Ghosh A. and Das S. K.: Coverage and connectivity issues in wireless sensor networks: A survey. Pervasive and Mobile Computing. **4(3)**, 303–334 (2008)

43. Goldsworthy A., *The Fall of Carthage: The Punic Wars 265-146 BC* (Orion Books, London, 2012)

44. Grinstead Ch. M. and Snell J. L., *Introduction to Probability* (American Mathematical Society, RI, 1997).

45. Harmati I. and Skrzypczyk K.: Robot team coordination for target tracking using fuzzy logic controller in game theoretic framework. Robotics and Autonomous Systems, **57**, 75–86 (2009)

46. Hayakawa T., Matsuzawa T. and Hara S.: Formation control of multi-agent systems with sampled information. Proceedings of IEEE Conference on Decision and Control. 4333–4338 (San Diego, CA, 2006)

47. Hazon N., Mieli F. and Kaminka G.: Towards robust on-line multi-robot coverage. IEEE International Conference on Robotics and Automation (ICRA). 1710–1715 (Edmonton Alberta, Canada, 2006)

48. Hong Y., Chen G. and Bushnell L.: Distributed observers design for leader-following control of multi-agent networks. Automatica. **44(3)**, 846–850 (2008)

49. Horn R. A. and Johnson C. R., *Matrix Analysis* (Cambridge University Press, Cambridge, U.K., 1985)

50. Hoy, M., Matveev, A. S. and Savkin, A. V.: Collision free cooperative navigation of multiple wheeled robots in unknown cluttered environments. Robotics and Autonomous Systems. **60(10)**, 1253–1266 (2012)

51. Hoy, M., Matveev, A. S. and Savkin, A. V.: Algorithms for collision free navigation of mobile robots in complex cluttered environments: A survey. Robotica. **33(3)**, 463-497 (2015)

52. Hsieh M. A., Kumar V. and Chaimowicz L.: Decentralized controllers for shape generation with robotic swarms. Robotica. **26(5)**, 691–701 (2008)

53. Jadbabaie A., Lin, J. and Morse A. S.: Coordination of groups of mobile autonomous agents using nearest neighbor rules. IEEE Transactions on Automatic Control. **48(6)**, 988–1001 (2003)

54. Jeremić A. and Nehorai A.: Design of chemical sensor arrays for monitoring disposal sites on the ocean floor. IEEE Journal of Oceanic Engineering. **23(4)**, 334–343 (1998)

55. Jiang F. and Wang L.: Finite-time information consensus for multi-agent systems with fixed and switching topologies. Physica D: Nonlinear Phenomena. **238(16)**, 1550–1560 (2009)

56. Kershner R.: The number of circles covering a set. American Journal of Mathematics. **61**, 665–671 (1939)

57. Killough S. M. and Pin F. G.: A new family of omnidirectional and holonomic wheel platforms for mobile robots. IEEE Transactions on Robotics and Automation. **10(4)**, 480–489 (1994)

58. Kloder S. and Hutchinson S.: Barrier coverage for variable bounded-range line-of-sight guards. Proceedings of the IEEE International Conference on Robotics and Automation. 391–396 (2007)

59. Kumar S., Lai T. H. and Arora A.: Barrier coverage with wireless sensors. Wireless Networks. **13(6)**, 817–834 (2007)

60. Kurabayashi D., Ota J., Arai T. and Yoshida E.: Cooperative sweeping by multiple mobile robots. Proceedings of the IEEE International Conference on Robotics and Automation. **2** 1744–1749 (1996)

61. Leonard N. E. and Fiorelli E.: Virtual leaders, artificial potentials and coordinated control of groups. Proceedings of the IEEE Conference on Decision and Control. **3** 2968–2973 (2001)

62. Li W. and Wang X.: Adaptive velocity strategy for swarm aggregation. Physical Review E: Statistical, Nonlinear, and Soft Matter Physics. **75(2)** (2007)

63. Li W., Zhang H. T., Chen M. Z. and Zhou T.: Singularities and symmetry breaking in swarms. Physical Review E: Statistical, Nonlinear, and Soft Matter Physics. **77(2)** (2008)

64. Lin F. Y. S. and Chiu P. L.: A near-optimal sensor placement algorithm to achieve complete coverage/discrimination in sensor networks. IEEE Communications Letters. **169(1)**, 43–45 (2005)

65. Liu B. and Towsley D.: A study of the coverage of large-scale sensor networks. IEEE International Conference on Mobile Ad-Hoc and Sensor Systems. 475-483 (2004)

66. Ma M. and Yang Y.: Adaptive triangular deployment algorithm for unattended mobile sensor networks. IEEE Transactions on Computers. **56(7)**, 946–958 (2007)

67. Manchester I. R. and Savkin A. V.: Circular navigation guidance law for precision missile/target engagement. Journal of Guidance, Control, and Dynamics. **29(2)**, 314–320 (2006)

68. Martinson E. and Payton D.: Lattice formation in mobile autonomous sensor arrays *Swarm Robotics, Lecture Notes in Computer Science.* 98–111 (Springer-Verlag, Berlin, 2005)

69. Mathews G. M., Durrant-Whyte H. and Prokopenko M.: Decentralised decision making in heterogeneous teams using anonymous optimisation. Robotics and Autonomous Systems. **57(3)**, 310–320 (2009)

70. Matveev A. S., Hoy M. and Savkin A. V.: A method for reactive navigation of nonholonomic robots in maze-like environments. Automatica. **49(5)**, 1268–1274 (2013)

71. Matveev A. S., Hoy M. and Savkin A. V.: A globally converging algorithm for reactive robot navigation among moving and deforming obstacles. Automatica. **54**, 292-304 (2015)

72. Matveev A. S. and Savkin A. V., *Estimation and Control over Communication Networks* (Birkhäuser, Boston, 2009)

73. Matveev A. S. and Savkin A. V.: The problem of state estimation via asynchronous communication channels with irregular transmission times. IEEE Transactions on Automatic Control. **48(4)**, 670–676 (2003)

74. Matveev A. S. and Savkin A. V.: An analogue of Shannon information theory for detection and stabilization of via noisy discrete communication channels. SIAM Journal on Control and Optimization. **46(4)**, 1323–1367 (2007)

75. Matveev A. S., Savkin A. V., Hoy M. and Wang C., *Safe Robot Navigation among Moving and Steady Obstacles* (Elsevier, 2015)

76. Matveev A. S., Teimoori H. and Savkin A. V.: A method for guidance and control of an autonomous vehicle in problems of border patrolling and obstacle avoidance. Automatica. **47(3)**, 515–524 (2011)

77. Matveev A. S., Teimoori H. and Savkin A. V.: Range-only measurements based target following for wheeled mobile robots. Automatica. **47(1)**, 177–184 (2011)

78. Matveev A. S., Teimoori H. and Savkin A. V.: Navigation of a unicycle-like mobile robot for environmental extremum seeking. Automatica. **47(1)**, 85–91 (2011)

79. Matveev A. S., Teimoori H. and Savkin A. V.: Method for tracking of environmental level sets by a unicycle-like vehicle. Automatica. **48(9)**, 2252–2261 (2012)

80. Matveev A. S., Wang C. and Savkin A. V.: Real-time navigation of mobile robots in problems of border patrolling and avoiding collisions with moving and deforming obstacles. Robotics and Autonomous Systems. **60(6)**, 769–788 (2012)

81. Meyer C. D., *Matrix Analysis and Applied Linear Algebra.* (SIAM, Philadelphia, 2000)

82. Min T. W. and Yin H. K.: A decentralized approach for cooperative sweeping by multiple mobile robots. Proceedings of IEEE/RSJ International Conference on Intelligent Robots and Systems. 380–385 (Victoria, BC , Canada, 1998)

83. Nazarzehi Had V., Savkin A. V. and Baranzadeh A.: Distributed 3D dynamic search coverage for mobile wireless sensor networks. IEEE Communications Letters. **to appear** (2015)

84. Okubo A.: Dynamical aspects of animal grouping: Swarms, schools, flocks, and herds. Advances in Biophysics. **22**, 1–94 (1986)

85. Olfati-Saber R., Fax J. A. and Murray R. M.: Consensus and cooperation in networked multi-agent systems. Proceedings of the IEEE. **95(1)**, 215–233 (2007)

86. Pathirana P. N., Bulusu N., Savkin A. V. and Jha S.: Node localization using mobile robots in delay-tolerant sensor networks. IEEE Transactions on Mobile Computing. **4(3)**, 285–296 (2005)

87. Pathirana P. N., Savkin A. V. and Jha S.: Location estimation and trajectory prediction for cellular networks with mobile base stations. IEEE Transactions on Vehicular Technology. **53(6)**, 1903–1913 (2004)

88. Petersen I. R. and Savkin A. V., *Robust Kalman Filtering for Signals and Systems with Large Uncertainties.* (Birkhauser, Boston, 1999)

89. Pettersen K. Y., Gravdahl J. T., and Nijmeijer H. Eds., *Group Coordination and Cooperative Control, Lecture Notes in Control and Information Sciences* (Springer, London, 2006)

90. Reif J. H. and Wang H.: Social potencial fields: A distributed behavioral control for autonomous robots. Robotics and Autonomous Systems. **27**, 171–194 (1999)

91. Rekleitis I., New A. P., Rankin, E. S. and Choset H.: Efficient boustrophedon multi-robot coverage: An algorithmic approach. Annals of Mathematics and Artificial Intelligence. **52(2–4)**, 109–142 (2008)

92. Ren W. and Beard R. W., *Distributed Consensus in Multi-Vehicle Cooperative Control* (Springer, London, 2008)

93. Reynolds C. W.: Flocks, birds, and schools: A distributed behavioral model. Computer Graphics. **21**, 25–34 (1986)

94. Royden H. L., *Real Analysis* (Prentice Hall, Englewood Cliffs, NJ, 1988)

95. Saffarian M. and Fahimi F.: Non-iterative nonlinear model predictive approach applied to the control of helicopters' group formation. Robotics and Autonomous Systems. **57(6–7)**, 749–757 (2009)

96. Savkin A. V.: Coordinated collective motion of groups of autonomous mobile robots: Analysis of Vicsek's model. IEEE Transactions on Automatic Control. **49(6)**, 981–983 (2004)

97. Savkin A. V.: Analysis and synthesis of networked control systems: Topological entropy, observability, robustness, and optimal control. Automatica. **42(1)**, 51–62 (2006)

98. Savkin A. V. and Cheng T. M.: Detectability and output feedback stabilizability of nonlinear networked control systems. IEEE Transactions on Automatic Control. **52(4)**, 730–735 (2007)

99. Savkin A. V. and Hoy M.: Reactive and shortest path navigation of a wheeled mobile robot in cluttered environments. Robotica. **31(2)**, 323–330 (2013)

100. Savkin A. V. and Javed F.: A method for decentralized self-deployment of a mobile sensor network with given regular geometric patterns. Proceedings of the Seventh International Conference on Intelligent Sensors, Sensor Networks and Information Processing (Adelaide, Australia, 2011)

101. Savkin A. V., Javed F. and Matveev A. S.: Optimal distributed blanket coverage self-deployment of mobile wireless sensor networks. IEEE Communications Letters. **16(6)**, 949–951 (2012)

102. Savkin A. V. and Petersen I. R.: Recursive state estimation for uncertain systems with an integral quadratic constraint. IEEE Transactions on Automatic Control. **40(6)**, 1080–1083 (1995)

103. Savkin A. V. and Petersen I. R.: Robust state estimation and model validation for discrete-time uncertain systems with deterministic description of noise and uncertainty. Automatica. **34(2)**, 271–274 (1998)

104. Savkin A. V. and Petersen I. R.: Set-valued state estimation via a limited capacity communication channel. IEEE Transactions on Automatic Control. **48(4)**, 676–680 (2003)

105. Savkin A. V. and Teimoori H.: Decentralized formation flocking and stabilization for networks of unicycles. Proceedings of the IEEE Conference on Decision and Control. 984–989 (Shanghai, China, 2009)

106. Savkin A. V. and Teimoori H.: Decentralized navigation of groups of wheeled mobile robots with limited communication. IEEE Transactions on Robotics. **26(6)**, 1099–1104 (2010)

107. Savkin A. V. and Wang C.: A simple biologically-inspired algorithm for collision free navigation of a unicycle-like robot in dynamic environments with moving obstacles. Robotica. **31(6)**, 993–1001 (2013)

108. Savkin A. V. and Wang C.: Seeking a path through the crowd: Robot navigation in unknown dynamic environments with moving obstacles based on an integrated environment representation. Robotics and Autonomous Systems. **62(10)**, 1568–1580 (2014)

109. Savkin A. V., Wang, C., Baranzadeh A., Xi Z. and Nguyen, H.T.: A method for de-centralized formation building for unicycle-like mobile robots. Proceedings of the 9th Asian Control Conference.(Istanbul, Turkey, 2013)

110. Savkin A. V., Xi Z. and Nguyen, H.T.: An algorithm of decentralized encircling cover-age and termination of a moving deformable region by mobile robotic sensor/actuator networks. Proceedings of the 9th Asian Control Conference (Istanbul, Turkey, 2013)

111. Shaw E.: The schooling of fishes. Scientific American. **206**, 128–138 (1962)

112. Shen C., Cheng W., Liao X. and Peng S.: Barrier coverage with mobile sensors. Pro-ceedings of the International Symposium on Parallel Architectures, Algorithms, and Networks. 99–104 (Sydney, Australia, 2008)

113. Shi G. and Hong Y.: Global target aggregation and state agreement of nonlinear multi-agent systems with switching topologies. Automatica. **45(5)**, 1165–1175 (2009)

114. Tan G., Jarvis S. and Kermarrec A. M.: Connectivity-guaranteed and obstacle-adaptive deployment schemes for mobile sensor networks. IEEE Transactions on Mobile Com-puting. **8(6)**, 836–848 (2009)

115. Tanner H. G. and Christodoulakis D. K.: Decentralized cooperative control of hetero-geneous vehicle groups. Robotics and Autonomous Systems. **55(11)**, 811–823 (2007)

116. Teimoori H. and Savkin A. V.: Equiangular navigation and guidance of a wheeled mobile robot based on range only measurements. Robotics and Autonomous Systems. **58(2)**, 203–215 (2010)

117. Teimoori H. and Savkin A. V.: A biologically inspired method for robot navigation in a cluttered environment. Robotica. **28(5)**, 637-648 (2010)

118. The United Nations Convention on the Law of the Sea (United Nations, 1982)

119. Tsitsiklis J.N.: Problems in Decentralized Decision Making and Computation, Ph.D. Thesis, MIT, Cambridge, MA (1984)

120. Vicsek T., Czirok A., Jacob E. Ben, Cohen I. and Schochet O.: Novel type of phase transitions in a system of self-driven particles. Physical Review Letters. **75**, 1226–1229 (1995)

121. Wagner I., Lindenbaum M. and Bruskstein A. M.: Distributed covering by ant-robots using evaporating traces. IEEE Transactions on Robotics and Automation. **15**, 918–933 (1999)

122. Wang G., Cao G. and La Porta T. F.: Movement-assisted sensor deployment. IEEE Transactions on Mobile Computing. **5(6)**, 640–652 (2006)

123. Wang P. K. C.: Navigation strategies for multiple autonomous mobile robots moving in formation. Journal of Robotic Systems. **8(2)**, 177–195 (1991)

124. Wang W., Srinivasan V. and Chua K.: Coverage in hybrid mobile sensor networks. IEEE Transactions on Mobile Computing. **7(11)**, 1374–1387 (2008)

125. Wang Y. and Tseng Y.: Distributed deployment schemes for mobile wireless sensor networks to ensure multilevel coverage. IEEE Transactions on Parallel and Distributed Systems. **19(9)**, 1280–1294 (2008)

126. Warburton K. and Lazarus J.: Tendency-distance models of social cohesion in animal groups. Journal of Theoretical Biology. **150(4)**, 473–488 (1991)

127. Wu Z., Guan Z., Wu X. and Li T.: Consensus based formation control and trajectory tracing of multi-agent robot systems. Journal of Intelligent and Robotic Systems. **48(3)**, 397–410 (2007)

128. Yamashita A., Arai T., Ota J., and Asama H.: Motion planning of multiple mobile robots for cooperative manipulation and transportation. IEEE Transactions on Robotics and Automation. **19(2)**, 223–237 (2003)

129. Yang G. and Qiao D.: Critical conditions for connected-k-coverage in sensor networks. IEEE Communications Letters. **12(9)**, 651–653 (2008)

130. Yang W., Cao L., Wang X. and Li X.: Consensus in a heterogeneous influence network. Physical Review E: Statistical, Nonlinear, and Soft Matter Physics. **74(3)** (2006)

131. Yu H. and Wang Y.: Coordinated collective motion of groups of autonomous mobile robots with directed interconnected topology. Journal of Intelligent and Robotic Systems. **53**, 87–98 (2008)

132. Zakharyeva, A., Matveev, A. S., Hoy, M., Savkin, A. V.: Distributed control of multiple non-holonomic robots with sector vision and range-only measurements for target capturing with collision avoidance. Robotica. **33(2)**, 385-412 (2015)

133. Zhai C. and Hong Y.: Decentralized sweep coverage algorithm for multi-agent systems with workload uncertainties. Automatica. **49(7)**, 2154–2159 (2013)

134. Zhang H. and Pathirana P. N.: Optimization-based formation of autonomous mobile robots. Robotica. **29(4)**, 515–525 (2011)

135. Zhang J., Zhao Y., Tian B., Peng L., Zhang H. -T., Wang B. H. and Zhou T.: Accelerating consensus of self-driven swarm via adaptive speed. Physica A: Statistical Mechanics and Its Applications. **388(7)**, 1237–1242 (2009)

136. Zheng X., Jain S., Koenig S. and Kempe D.: Multi-robot forest coverage. IEEE International Conference on Intelligent Robots and Systems (IROS). **2**, 3852–3857 (Edmonton Alberta, Canada, 2005)

137. Zhu S., Wang D. and Low C. B.: Cooperative control of multiple UAVs for moving source seeking. Journal of Intelligent and Robotic Systems. **74**, 333–346 (2014)

Index

Decentralized Coverage Control Problems for Mobile Robotic Sensor and Actuator Networks. **191**
By Andrey V. Savkin, Teddy M. Cheng, Zhiyu Xi, Faizan Javed, Alexey S. Matveev, and Hung
Nguyen. Copyright © 2015 by the Institute of Electrical and Electronics Engineers, Inc.

IEEE PRESS SERIES ON SYSTEMS SCIENCE AND ENGINEERING

Editor:
MengChu Zhou, *New Jersey Institute of Technology and Tongji University*

Co-Editors:
Han-Xiong Li, *City University of Hong-Kong*
Margot Weijnen, *Delft University of Technology*

The focus of this series is to introduce the advances in theory and applications of systems science and engineering to industrial practitioners, researchers, and students. This series seeks to foster system-of-systems multidisciplinary theory and tools to satisfy the needs of the industrial and academic areas to model, analyze, design, optimize and operate increasingly complex man-made systems ranging from control systems, computer systems, discrete event systems, information systems, networked systems, production systems, robotic systems, service systems, and transportation systems to Internet, sensor networks, smart grid, social network, sustainable infrastructure, and systems biology.

1. *Reinforcement and Systemic Machine Learning for Decision Making*
   Parag Kulkarni
2. *Remote Sensing and Actuation Using Unmanned Vehicles*
   Haiyang Chao and YangQuan Chen
3. *Hybrid Control and Motion Planning of Dynamical Legged Locomotion*
   Nasser Sadati, Guy A. Dumont, Kaveh Akbari Hamed, and William A. Gruver
4. *Modern Machine Learning: Techniques and Their Applications in Cartoon Animation Research*
   Jun Yu and Dachen Tao
5. *Design of Business and Scientific Workflows: A Web Service-Oriented Approach*
   Wei Tan and MengChu Zhou
6. *Operator-based Nonlinear Control Systems: Design and Applications*
   Mingcong Deng
7. *System Design and Control Integration for Advanced Manufacturing*
   Han-Xiong Li and XinJiang Lu
8. *Sustainable Solid Waste Management: A Systems Engineering Approach*
   Ni-Bin Chang and Ana Pires
9. *Contemporary Issues in Systems Science and Engineering*
   Mengchu Zhou, Han-Xiong Li, and Margot Weijnen
10. *Decentralized Coverage Control Problems For Mobile Robotic Sensor and Actuator Networks*
    Andrey V. Savkin, Teddy M. Cheng, Zhiyu Li, Faizan Javed, Alexey S. Matveev, and Hung Nguyen